Glossary of Milling and Baking Terms

First Edition

Other books by the same author—

Chemistry and Technology of Cereals as Food and Feed
First and Second Editions

Bakery Technology and Engineering
First and Second Editions

Equipment for Bakers

Cereal Technology

Cereal Science

Food Texture

Water in Foods

Bakery Technology

Ingredients for Bakers

Cookie and Cracker Technology
First, Second, and Third Editions

Formulas and Processes for Bakers

Technology of the Materials of Baking

Snack Food Technology
First, Second, and Third Editions; Japanese Edition

* * * * * *

This book was designed and manufactured to give you maximum value and long use. It is printed on good quality, heavy base-weight paper of superior opacity and whiteness, so that it will last longer and be easier to read. The type face is New Century Schoolbook, an open style which is esthetically pleasing and can be scanned quickly and accurately.

The book has been bound by a process called "Smyth sewn in signatures," a method selected to give a sturdy book with convenient handling properties. The binding was made of stronger-than-usual papers and boards to support the pages without sagging. The cover is plastic impregnated cloth which is resistant to wear, stains, and other damage; its gold stamping helps to make the book an attractive addition to your library.

GLOSSARY OF MILLING AND BAKING TERMS

by

SAMUEL A. MATZ, PH. D.

President, Pan-Tech International, Inc. Formerly, Vice President for Research, Development, and Compliance, Ovaltine Products, Inc. At one time, Vice President for Research and Development, Robert A. Johnston Co.; Technical Director of the Refrigerated Dough Program, Borden Foods Co.; Chief of the Cereal and General Products Branch, Quartermaster Food and Container Institute for the Armed Forces; Chief Chemist, Harvest Queen Mill and Elevator Co.; Instructor, Department of Flour and Feed Milling Industries, Kansas State University; Chemist, Iglehart Mills Division of General Foods.

PAN-TECH INTERNATIONAL, INC.
P. O. BOX 4548
MC ALLEN, TEXAS 78502
1993

Copyright 1993 by SAMUEL A. MATZ

ISBN 0-942849-26-4

All rights reserved. No part of this work covered by the copyright hereon may be reproduced or used in any form or by any means — graphic, electronic, or mechanical, including photocopying, recording, taping, or information storage and retrieval systems — without written permission of the copyright owner.

Library of Congress Cataloging-in-Publication Data:

Matz, Samuel A.
    Glossary of milling and baking terms / by Samuel A. Matz. -- 1st ed.

      p.   cm.
  ISBN 0-942849-26-4
    1. Grain--Milling--Dictionaries. 2. Baking--Dictionaries.
I. Title.
TX763.M337   1993
664'.720'03--dc20   93-8685
                                                            CIP

PRINTED IN THE UNITED STATES OF AMERICA

PREFACE

According to the unabridged third edition of Merriam-Webster's "Webster's New International Dictionary," a **Glossary** is, "A collection of glosses, or explanations of words and passages of a work or author; a partial dictionary of words and passages of a work, an author, a dialect, art or science, explaining technical terms or uncommon words." The book you are reading is not intended to be an all-purpose or all-inclusive dictionary. It contains definitions of many words and phrases that may be encountered by students and workers in publications on milling, baking, and other fields of cereal technology. The definitions are specifically applicable to the texts of the several books on bakery technology that I have written. Processes, machinery, utensils, tests, ingredients, and products are the primary areas from which defined words have been selected.

The words defined in the following pages have placed in alphabetical order, with spaces, hyphens, apostrophes, accent markings, and capitalization ignored in their positioning. This is not in accord with the practices in many dictionaries and indexes, but it seemed to me to be the most helpful alternative for the present purpose and, for that matter, the most logical method for alphabetically organizing any list of words.

The extensive collection of terms defined in this glossary includes numerous foreign language entries. The principal geographic area of use (if other than the U.S.A.) is denoted by a word or abbreviation in italics immediately following the defined word. The abbreviations used here are: *Ch*=China, *Fr*=France, *Ger*=Germany, *It*=Italy, *Ind*=India, *Jap*=Japan, *ME*=Middle East in the broadest sense, *Mex*=Mexico (often but not always applicable to Latin America in general), *Sp*=Spain (sometimes but not always applicable to Latin America as well), and *UK*=England. Other geographical areas — primarily nations — are spelled out. If a word or phrase has become widely accepted in U.S.A. cereal and bakery circles, no indication of linguistic origin has been given.

Samuel A. Matz
Edinburg, Texas
July 1, 1993

GLOSSARY OF MILLING AND BAKING TERMS

-A-

absorption — (1) The amount of water required to be added to a particular flour in order for it to function optimally in some application (usually the amount required to make the best possible bread dough), expressed as a percent of the flour weight. (2) Amount of oil or fat retained by a food product that has been fried.

accelerated rancidity test — any method of estimating the storage life of fats or fat-containing products by speeding up the onset and progress of rancidity in a sample of the material. Usually involves increasing the temperature, oxygen tension, light intensity, etc.

aceite — *Sp* oil.

acetic acid — an organic acid, CH_3COOH. The pure substance (glacial acetic acid) is a colorless, pungent liquid that congeals at cool room temperatures. Acetic acid of commerce is an aqueous solution containing various percentages of the compound. Vinegar contains 4.5 to 12% (typically 5%) acetic acid. It has long been used as a preventative of rope spoilage in bread.

acid — variously defined in different times and schools, but, in the broadest chemical sense, any substance capable of donating protons to bases.

acid conversion method — a hydrolysis process for breaking down corn starch into glucose and other saccharides in the form of corn syrup.

acidic — (1) Having a pH of less than 7. (2) Having a tart or sour taste.

acidulant — a substance added to a system such as a beverage in order to increase the acid taste and/or reaction of the system.

Acid Value — a measure of the amount of free fatty acids in a sample of fatty material, reported as the number of milligrams of potassium hydroxide required to neutralize the free fatty acids in a gram of fat.

acrid — harsh or bitterly pungent in taste or aroma.

Activated Dough Development — according to the Flour Milling and Baking Research Association of the UK, a process of breadmaking in which no time has been allocated for bulk fermentation; adequate development of the dough is achieved by adding fairly large amounts of cysteine, ascorbic acid, and potassium bromate.

active dry yeast — a preparation of live bakers' yeast that has been dried to about 6% to 8% moisture; usually offered in the form of granules or small pellets. Preservatives and other additives may be present.

Active Oxygen Method — an accelerated rancidity test in which the fat to be tested is held at 98°C while air is bubbled through it at a specified rate. The endpoint is reported as hours needed to reach a peroxide value of 100 meq/Kg. There is a variable relationship between AOM hours of an oil and the actual shelf-life of a product containing the oil.

additive — according to the FDA: any substance, the intended use of which results or may reasonably be expected to result, directly or indirectly, in its becoming a component of, or otherwise affecting, the characteristics of any food.

adulterate — to clandestinely add some foreign or inferior substance to a food or ingredient, usually for the purpose of making the food material cheaper to produce, as added water would be an adulterant in milk.

aeration — the leavening of a dough or (more often) a batter by mixing in air or (less often) by injecting compressed gas. Sometimes applied to any kind of leavening action.

aerobic — applied to microorganisms that require uncombined oxygen (O_2) for growth; contrasted with anaerobic organisms that grow best in atmospheres of very low oxygen content.

aflatoxin — one of a related group of organic compounds produced by certain *Aspergillus* species; some are extremely poisonous.

agar — (agar agar) powder derived from sea plant extractives; forms a very firm gel with large percentages of water. Used in decorating jellies, etc.

ägg — *Sweden* egg.

ägghvitor — *Sweden* egg whites.

äggular — *Sweden* egg yolks.

aging — a step in the milling process in which flour is stored for a considerable time after grinding so that its original creamy tint is greatly reduced by natural reactions and its baking quality is improved. Seldom, if ever, used in modern mills in the U.S.A.

agitator — the part (paddle, whip, dough hook, etc.) of a mixer that causes turbulence in the bowl's contents.

agua — *Sp* water.

air-belt purifier — a purifier (q.v.) in which the same air is recirculated continuously through the sieve, fan, and dust collector.

air classification — a process by which the particles in a mill stream are separated according to size and density by cyclone separators instead of sieves.

ajo — *Sp* garlic.

akhroot — *Ind* walnut.

albumen — egg white.

albumin — a type of protein, characterized by its solubility and chemical properties.

alcohol — in general, an organic chemical compound containing the hydroxyl group -OH. Specifically ethanol (grain alcohol).

alemares — *Mex* pretzel shapes made from pan fino dough and sprinkled with coarse sugar before baking.

alginate — salts of alginic acid found in certain kinds of algae and used as viscosity improvers in foods and beverages.

alimento — *Sp* food.

allspice — the dried, nearly ripe fruit of *Pimenta dioica*. Has a strong clove note with nuances of cinnamon and nutmeg.

almendra — *Sp* almond.

almibar — *Sp* sirup.

almond — a nut much favored for use in and on bakery products, the kernel or seed of a small tree. Available in many sizes and varieties.

almond paste — almonds ground with sugar to yield a stiff paste; sometimes a binding ingredient is added. Used as an ingredient in bakery fillings, etc.

aloo — *Ind* potato.

alum — several aluminum sulfate salts have been called alum, but for the baker it nearly always means sodium aluminum sulfate, which is used as the acid reacting component of some baking powders.

aluminum — a light relatively soft metal of silvery color. It is a good conductor of heat and electricity. When exposed to air and moisture, it quickly becomes covered with a thin hard coating of oxide that resists further corrosion.

alveograph — an instrument that measures the stretchability of dough; said to provide an indication of baking quality. A standard disc of dough is blown into a bubble and the pressure change and bursting pressure charted vs. time.

amaranth — a herbaceous plant that produces small edible seeds on a sorghum-like head; it is a grain but not a cereal. The U.S. crop is quite small, but the grain is an important food in some Asian and African countries. Has been used as a "grain" ingredient in multigrain breads.

amaretti — *It* macaroon cookies with a porous structure; baked dry and crisp. Amarettini are smaller cookies of the same general type.

ambient — surrounding or encompassing, as ambient conditions.

Amflow Process — a trade name used by the AMF Company for their version of continuous breadmaking plants.

amino acids — organic compounds containing both amino and carboxyl groups attached to a carbon chain. Proteins are made up of many of these groups joined together. Some of the amino acids are essential nutrients for humans.

ammonia — incorrect name for the ammonium bicarbonate-ammonium carbonate mixture sometimes used for leavening cookies, etc. Ammonia gas is never used as an ingredient.

ammonium chloride — a chemical compound sometimes used in yeast foods to provide a source of nitrogen for yeast growth.

ammonium phosphate — a nitrogen source for yeast growth used in dough improvers and in nutrient broths.

amylase — an enzyme that can hydrolyze starches.

amylograph — an instrument used to determine change of viscosity with time in a heated mixture of water and a starchy material such as flour.
amylolytic enzymes — those enzymes that hydrolyze starches and similar glucose polymers.
amylopectin — starch molecules made up of glucose units chemically combined in branching chains.
amylose — starch molecules made up of glucose units chemically combined in long unbranched chains.
ananás — *Sp* pineapples, in some localities this name is restricted to a smaller, sweeter type of pineapple, the others being called "piñas."
anaerobic — indicates an organism (anaerobe) that can grow and reproduce in environments having very low oxygen tension. Obligate anaerobes are actually inhibited by significant concentrations of O_2.
andruty — *Poland* wafer cookies.
angeer — *Ind* figs.
angoor — *Ind* grapes.
angel food cake — a highly aerated (low density) white cake made principally from whipped egg whites, flour, and sugar. Never contains shortening.
angel food pan — pans in which angel food cakes are baked, usually round with high, outward-sloping sides; they have a vertical cylindrical tube in the middle.
angelica — a herbaceous plant the leaves and stems of which are candied and used as decorations on cakes, etc. Not at all common in U.S.A. practice. Various parts of the plant have been used to flavor liqueurs.
anydroglucose unit — the $C_6H_{10}O_5$ unit that is bound to other such units to make up the starch molecule.
anhydrous — not containing moisture.
anise — the dried fruit of *Pimpinella anisum*. Has a flavor often described as "like licorice." In the bakery, this spice is used mostly in cookies.
anisidine value — an indication of the extent of oxidative deterioration of fats and oils based on spectrophotometric measurements of the aldehydes present in the lipid.
annatto — a food coloring preparation made by extracting the coating of the seed of *Bixa orellanna* trees. Yellowish to reddish-orange, and oil soluble.
antimicrobial agent — an additive that deters growth of microorganisms.
antioxidant — a substance that will reduce the rate at which a fat or oil oxidizes and becomes rancid.
apprêt — *Fr* proofing.
apprêt sur couche — *Fr* placing pieces of dough between the folds of a cloth for proofing.
aquaculture — the commercial culture of sea, lake, and river foodstuffs such as fish, oysters, and seaweed.

aqueous — containing water.
arancini — candied orange peel in strips.
arap pidesi — *Turkey* a pocket bread, like pita. Yeast-leavened.
arepa — *Sp/Venezuela* flat bread made of masa and processed like a tortilla, except thicker.
aromatic compounds — (1) Chemically, compounds that contain at least one benzene ring in their structure. (2) Compounds that have a fragrance or smell, usually restricted to those with pleasant odors.
aroo — *Ind* peach.
arrack — a liquor distilled from fermented palm sap, molasses, or rice mash.
arrowroot — a starchy preparation obtained from certain tropical plants and used as an ingredient for thickening sauces and fillings. Not a common ingredient in modern U.S.A. baked goods.
arroz — *Sp* rice.
asbestos — a mineral found in fibrous form, valuable because of its resistance to flame and its low heat conductance. Now in disrepute because of alleged bad effects on health.
ascorbic acid — vitamin C; occurs naturally in many foods but is also made synthetically in large quantities for use as a nutritional supplement and as an improving agent in yeast doughs.
aseptic — describes a closed system, as a beverage in a sealed bottle, that has been rendered essentially free of microorganisms by heating, radiation, or other means.
ash — the inorganic material left after flour (or other organic material) is burned. Frequently included in flour specifications as a criterion of the extraction rate.
aspartame — a synthetic nutritional sweetener about 100 times as sweet as sucrose, sold in the form of a white, water-soluble powder.
Aspergillus — a genus of common fungi causing spoilage of bakery products and other moist foods; their bluish or greenish mycelia are frequently found on outdated bread.
aspic — a jelly, usually made from gelatin and broth, in which meat, fish, fruits, etc. are suspended. Sometimes used as a component of meat pies or fruit covered torts, etc.
aspiration — use of controlled velocities of directional air streams to separate particles having different resistances to air flow.
aspirator — a grain cleaning apparatus that utilizes the separating power of air currents to remove low density impurities such as dust, light chaff, and bran particles from grains or other granular material.
atmospheric pressure — force exerted on an object by the air surrounding it; at sea leavel. Decreases with increasing altitude and fluctuates with changing atmospheric conditions.

atta — *Ind* whole wheat flour.
autolysis — self-digestion of cells, noticeable in compressed yeast that has been stored too long or at elevated temperatures, when the mass becomes brownish in color, semi-liquid in consistency, and unpleasant in odor.
avellana — *Sp* hazelnuts, filberts.
azodicarbonamide — a synthetic improving or maturing agent that strengthens dough by facilitating the linking of gluten molecules.
azúcar — *Sp* sugar.

-B-

baba — a dessert cake made from rich yeast dough, often served after it has been soaked with a sweetened rum-flavored syrup.

babki — *Poland* pastries.

Bacillus mesentericus — the bacteria responsible for spoilage of the rope type in bakery products. Their spores often survive the heat they encounter inside bread loaves during baking.

backbone — an obsolete term for the principal component of a flour blend.

backpulver — *Ger* baking powder.

bacterial amylase — an enzyme derived by extracting certain kinds of proteins from suitable types of bacterial cultures; differs from many other amylases in that it retains a large part of its activity even after being subjected to high temperatures.

badaam — *Ind* almonds.

bademli empare — *Turkey* sweet almond cakes/cookies; chemically leavened; coated with a sweet syrup after baking.

bademli kurabiye — *Turkey* small, crescent-shaped cookies with a high content of finely ground almonds.

badem tatlisi — *Turkey* a light, spongy almond cake moistened with sweet syrup. High in egg content.

baffle — an obstruction to the flow of gases or liquids. For example, baffles are used to direct the flow of hot gases in ovens, promoting either the uniformity or non-uniformity (zones) of temperatures inside the baking chamber.

bagel — a doughnut-shaped, yeast-fermented bread product, that has a characteristic brown smooth surface resulting from its immersion in hot water shortly before baking and usually has a dense tough crumb.

bagger — a machine that inserts any bakery product into a plastic or paper bag; generally refers to a bread bagging machine.

bagging out — depositing cookie dough, icing, or other paste out of a pastry bag.

baguette — *Fr* the most common bread loaf shape in France. Typically, a long (about 2 ft.) narrow, approximately cylindrical loaf weighing between 12 and 24 oz.

bain marie — a double boiler used to keep mixtures hot but below boiling temperatures.

baiser — *Fr* a baked meringue.

bajra — *Ind* millet.

bake — cook by dry heat in an oven or similar device.

baked Alaska — a mound of ice cream on a layer of cake, both covered with a thick layer of meringue and baked until the meringue is slightly browned.

bake-off format — a production process in which at least 95% of a bakery's output is made from frozen dough that is either purchased or made in the bakery's own plant.

bake-out — the amount of weight loss undergone by a dough or batter during its passage through the oven.

bakers' chocolate — roasted and finely ground cocoa bean nibs without additives; synonymous with chocolate liquor and bitter chocolate.

bakers' flour — a name formerly applied to first clears.

bakers' percent — the weight of individual ingredients expressed as a percentage of the weight of flour in the formula. Thus, a dough made from 100 lbs flour and 60 lbs water would have 60% ingredient water (60% absorption).

bakers' yeast — a preparation consisting mostly of cells of *Saccharomyces cerevisiae*.

baking powder — a mixture of sodium bicarbonate and an acid reacting substance, with other relatively inert ingredients. Used widely as a leavener in both home and commercial baking.

baking powder biscuit — chemically-leavened hot bread made in individual portion size, usually as flat round cakes. The dough is lean, and the finished product has a soft white crumb and a light brown crust on the top but not on the sides. This product can be found in many types that vary in shape, size, density, and other characteristics.

baking sheets — flat metal sheets on which dough pieces can be baked. They are found in a wide range of sizes, sometimes with low-rise rims, and can be constructed of iron, steel, or aluminum.

baking soda — sodium bicarbonate; available in several different particle sizes.

baklava — *Greece* a crisp confection of baked filo leaves enclosing chopped nuts; baked in a large square about 1 inch thick. A sweet syrup is poured over the baked product, then it is cut into pieces typically about 1 x 2 inches in size.

bakoom — *Egypt* disk-shaped, two-layered flatbread made from white flour, water, and salt. Proofed about 90 min before baking.

bakverk — *Sweden* pastry.

balady bread — one of several middle eastern pocket breads made of sourdough; usual piece size about 150 g.

balancing — in flour milling, adjusting the operations of the various breaks and separations in a roller mill to efficiently produce a desired product. In bakery product formulation, adjusting the proportions of ingredients so as to yield a workable dough and a quality product.

balloon whip — an agitator for an electric mixer that consists of many wires leading from a top disc around the bottom and back to the top on the other side. Used for whipping egg whites and the like.

GLOSSARY OF MILLING AND BAKING TERMS

band — (1) A paper or plastic strip placed around a loaf of bread or other bakery product as part of the package. (2) A metal hoop that is lined with paper and placed on a baking sheet to be filled with batter. (3) The moving continuous metal strip that forms the baking surface of a band oven.

band oven — an oven in which the hearth is a continuous belt of steel traveling around large cylinders at each end of the tunnel that constitutes the baking chamber.

band slicer — equipment for slicing bread loaves and the like; it has saw-like metal bands supported on two rotating drums.

banneton — a woven wicker basket in which a yeast-leavened dough is placed to ferment.

bannock — *UK/Scotland* flatbread, usually disk-shaped and unleavened. Cooked on a griddle. Variants have been made of oatmeal, barley flour, or wheat flour.

baps — the "breakfast rolls of Scotland." Doubtless many versions of this yeast-leavened bun can be found; there are recipes calling for just flour, water, salt, and yeast while others specify eggs, lard, milk, and other enrichments. "Floury baps" are said to be traditional; the dough pieces are brushed with milk (or water), then dusted with flour (this may be repeated) immediately before they are placed in the oven. The dough pieces are docked in one place with the forefinger before baking.

bara brith — *Welsh* a yeast-leavened, rich formula sweet bread, containing currants and/or raisins, candied fruit peels, etc. Flavored with various spices, such as cinnamon and nutmeg.

barbari — *Iran* a Middle Eastern flatbread, having an elongated oval shape with longitudinal ridges on top; it is washed with oil, etc., and extensively docked before baking. The average loaf is about an inch thick and weighs about 700 g.

barley — a cereal grain used for making malt and as a source of carbohydrates in brewing. It cannot be used as the main cereal ingredient for leavened bread because it does not form extensible doughs.

barm — an English/Australian term for the old-style starter cultures and sourdoughs formerly used to leaven bread; the predominant component is yeast. In olden days, barm was often obtained from breweries as foam collected from beer vats.

barometer — an instrument for determining atmospheric pressure.

barquettes — small boat-shaped pastries with either sweet or savory fillings.

bars — cookies having oblong or rectangular shapes. Made commercially by extruding dough through a bar press (rout press) machine.

base — (1) Bottom portion of a loaf of bread or other object. (2) A chemical compound that yields hydroxyl ions when dissolved in water; approximately the same as an alkali.

base mix — a preblend of the minor ingredients in a formula; used to improve scaling efficiency and reduce errors at the mixer. The base mix is added to the flour, water, and other major ingredients when the doughs are being prepared.

basmati — *Ind* a type of long-grain rice regarded as being superior in its culinary properties to other types of long-grain rice.

bassinage — *Fr* making a dough softer by adding more water during the kneading step.

bâtard — *Fr* a shorter, thicker loaf than the baguette. Typically, a foot in length and 24 oz. in weight.

batch — the amount of dough, etc., produced at one mixing.

Bath bun — — *UK* a light textured sweet roll, generally round, and usually containing currants. Its name refers to the city in England where it reputedly originated.

battawi — *Egypt* disk-shaped flatbread made from high extraction wheat flour, water, and salt. Dough baked immediately after mixing. Said to usually include 2 or 3 percent of fenugreek seed.

batter — a homogeneous flowable mass prepared by mixing flour and other ingredients (including water or other liquid ingredients).

batter beater — an agitator for an electric mixer; in shape it is a flat paddle having about three thick arms leading from each side of the top to a central post. It is a versatile agitator, but is most often used for cake batters.

batter sponge — a very soft pumpable ferment or sponge-phase mixture used in some breadmaking processes.

Baumé — a measure of the soluble solids in a sugar syrup, expressed as "degrees Baumé." Often spelled (incorrectly) Beaumé.

bazlama — *Turkey* flat bread made from flour, water, salt, and sourdough. Dough fermented 2 to 3 hr, cut into pieces of about 0.5 lb weight, flattened, and baked immediately on a hot griddle.

beard — the small bristles that grow at one end of the wheat kernel.

beat — to incorporate air into a batter by using rapid vigorous strokes of the agitator.

beaten biscuit — also, "Maryland biscuit." a kind of small cracker made of a lean dough that has been leavened by repeated folding and pounding.

beeswing — the two outer layers of the bran coat of the wheat berry, so-called because of their cross markings and light texture.

beignets — *Fr* a pastry made with pâte à chou and deep fried.

bench — the table on which the baker manipulates his dough.

bench brush — a brush about twelve inches long, used for cleaning the bench.

bench tolerance — the length of time a dough will retain good processing capability after it has been mixed and before it is baked.

benne seeds — sesame seeds.
benzoic acid — a food preservative generally effective only in acidic environments; not much used in bakeries.
benzoyl peroxide — a common bleaching agent used by millers to treat flour intended for yeast-leavened products.
berliini pannkoogid — *Estonia* doughnuts.
berry — (1) A kernel of wheat, and sometimes other cereal grains as well. (2) One of several kinds of small roundish juicy fruits without stones, e.g., blackberries and raspberries.
besan — *Ind* gram flour, chickpea flour.
beta amylase — enzyme that hydrolyzes soluble starch and dextrin to produce maltose; known as the saccharifying enzyme.
betabel — *Sp* beet.
BHA — butylated hydroxanisole, a synthetically prepared substance that retards the oxidation of fats, thereby slowing the development of rancidity.
bhajia — *Ind* savory fritters.
BHT — a chemical antioxidant very similar in its action to BHA.
bialy — a yeast-leavened roll flavored with onion bits and/or garlic, often made in the shape of a bagel, at other times formed into a flat round bun with a central circular depression.
biaxially oriented film — a plastic film (such as polypropylene) that has had its molecular structure highly oriented in two dimensions; this treatment improves many of the film's physical properties, including transparency and strength.
bimbollos — *Mex* a trade name for hamburger buns with sesame seeds
bimbuñuelos — *Mex* trade-named version of buñuelos, consisting of a fried, octagonal sweet bread with a topping of powdered sugar.
bin — a kind of box used for holding dry ingredients; found in a wide range of sizes and in many shapes.
biscotel — *Mex* cinnamon roll-shaped pastry topped with powdered sugar and pecans.
biscotte — *Fr* a rusk made by toasting slices of bread made from a fairly rich dough, yeast-leavened.
biscuit — in the U.S.A., a small round soft-textured bread made of a fairly lean dough leavened with baking powder; a baking powder biscuit. In most other English-speaking nations, "biscuits" refers to cookies and crackers, i.e., small dough pieces, either sweet or savory, baked to a low moisture content.
biskviit rabarbriga — *Estonia* like a pineapple upside-down cake, but made with rhubarb.
bitter chocolate — cacao nibs that have been roasted and finely ground; synonyms are bakers' chocolate and chocolate liquor.
bizcocho — *Sp* sponge cake or similar type of cake or bun.

black bun — *Scotland* a cake formed from a butter enriched yeast-leavened dough, fairly sweet and containing dried fruits, nuts, etc., that has been wrapped in a moderately thick (perhaps 0.2 to 0.5 inch) layer of a dough not containing fruits and nuts. The loaf may weigh as much as 5 or 6 lbs. Placed in a hoop for baking.

blackstrap — a kind of molasses having a very dark color, a strong bitter flavor, and a relatively low sugar content. Used mainly for animal feeds.

blanch — to briefly immerse in boiling water or in some other way partially cook the surface of a piece of food. Most blanched nuts have undergone a process of this type before being "skinned."

blancmange — a dessert made of milk, gelatine, sugar, starch, flavor, and color. Sometimes with fruit pieces suspended in it. Usually molded in fancy shapes.

blancs d'oeuf — *Fr* egg whites.

bland — having little or no flavor; insipid.

blanquillo — *Sp* egg whites.

blast freezer — a room or other inclosed space through which a high velocity current of cold (about -30°F) air is forced so as to quickly freeze the food products contained therein.

blé — *Fr* wheat.

blé sarrasin — *Fr* buckwheat.

bleached flour — wheat flour that has been treated with chemicals such as benzoyl peroxide in order to increase its whiteness; usually, some maturing action is also exerted by these substances.

bleaching — (1) Treating flour with oxidizing agents to decolorize some of the natural pigments and/or cause desirable changes in the gluten proteins. (2) In oil processing, a treatment to remove natural pigments and other impurities; the colored substances are commonly absorbed on activated charcoal or, more commonly, bleaching earth or clay.

bleeding — losing gas from cut dough edges.

blending — (1) The process of combining lots of wheat or other grain from different bins into a uniform batch. (2) Making a wet or dry mixture of ingredients, not necessarily either a dough or batter.

blind — (1) A condition in which the spaces in a sieve or screen have been clogged with sifted material so that particles cannot pass through. (2) The unintentional and undesirable sealing together of patterns cut or impressed on a piece of cookie or other dough, so that the desired design does not appear on the baked product. (3) Describes a pastry shell that has been baked without a filling.

blini — a Russian version of pancakes made, usually, with yeast and buckwheat flour, and eaten with sour cream, caviar, etc.

blister — a hollow bump with a cavity beneath it that has formed on the surface of a dough piece during baking or frying.

GLOSSARY OF MILLING AND BAKING TERMS

bloom — (1) The desirable visual texture and bright color found on the crust of well-baked loaves and rolls. (2) A defect in the appearance of chocolate consisting of a dull gray or white surface coating; it results either from moisture contacting the surface or exposure of the piece to temperatures near or above the melting point.

bocadillo — *Venezuela* a rich fruit cake made from crystallized fruits and rinds, sold in packets wrapped in banana leaves.

BOD — biological oxygen demand; the total amount of oxygen taken up by the micro organisms in a sample of water maintained at specified conditions.

body — (1) Response of the crumb to pressure; one aspect of bread texture. (2) Consistency or viscosity of a plastic or semifluid mixture, such as a starch pudding.

boil — to add enough heat to a liquid material so that some of the liquid is at all times turning into vapor and escaping from the surface as bubbles.

boiled icing — an icing made by boiling sugar and water to thread stage (238°F) then slowly adding it to beaten egg whites with additional beating.

boiled meringue — a meringue made by pouring sugar syrup boiled to the hard ball stage (about 250°F) over beaten egg white. Also called Italian meringue.

boiler — a vessel designed to furnish a supply of pressurized steam by efficiently and continuously transferring heat to liquid water.

bolillo — *Sp* small (about 2 to 4 oz) football-shaped, yeast-leavened hard or soft bread roll.

bollo — *Sp* biscuit or small cake.

bolting — sifting, esp., particle classification in a flour mill.

bombe — a molded (usually dome-shaped) dessert constructed from several different types of ice cream. Cut in portions immediately before serving, and often topped with a sauce.

bonbon — a small candy piece, usually a soft center coated with chocolate or similar enrobing material.

börek — *Turkey* pastries made out of thin layers of dough and butter, with various kinds of cheese or meat filling.

Boston brown bread — a dark sweet bread (chemically-leavened) containing cornmeal and molasses, and often raisins and spices, among other ingredients. Traditionally steamed, not baked.

Boston cream pie — two cake layers (usually white) separated by a thick layer of vanilla pastry cream (or starch pudding) and frosted with chocolate icing.

bottom heat — in an oven, the heat that is transferred to the product (or the pan) from the hearth.

botulism — acute food poisoning caused by a toxic chemical originating from *Clostridium botulinum*; often fatal.

bouchée — *Fr* a shell of puff pastry for one serving, such as a patty shell.
boulanger — *Fr* baker, esp., a baker of bread-type products.
boulangerie — *Fr* bakery.
bound water — water molecules that have their water-like properties chemically or physically reduced by other substances in their environment.
bowl extension — a metal or plastic ring or hood placed at the top of a mixer bowl so as to prevent material from being thrown out of the bowl during violent agitation.
bowl knife — a plastic spatula or flexible dull-edged knife used to scrape batter or dough from the sides and bottom of a mixer bowl.
brake — to pass through a dough brake; a dough brake (q.v.).
bran — in the milling of grain, the fraction consisting mostly of the fibrous outer layers of the kernels.
brandy — an alcoholic liquor distilled from wine, sometimes used as a flavoring agent in bakery foods or their adjuncts.
bran muffin — a chemically leavened quick bread in cupcake shape and containing some bran; usually colored dark brown with caramel coloring and often contains raisins.
bread — in a narrow sense, baked foods made of a developed dough containing flour, water, yeast, and salt, and usually containing malt, shortening, and milk. May contain other ingredients as well. The dough is fermented and otherwise processed before baking to give low density loaf or roll with a clearly defined crust and a soft silky crumb. See also, specific kinds of bread, such as French bread.
bread crumbs — the small particles generated when a loaf is sliced; also, ground dried bread used as a coating for fried foods.
bread faults — ways in which a specific loaf deviates from predetermined specifications and standards.
bread flour — generally, a hard wheat flour of at least moderately high protein content that has been milled and blended so as to make it particularly suitable for processing into bread.
bread improver — any of the various compounds and mixtures thereof that compensate for some of the inadequacies of a less-than-optimum flour used in breadmaking.
bread scoring — a system of evaluating bread that consists of applying numerical scores to various quality features of the loaf, and summing these individual scores to give an overall single-figure characterization of the quality of the product.
break — (1) One of the first steps by which the grain is reduced to meal in roller milling processes; usually performed by pairs of grooved steel cylinders. (2) The portion of inner crust exposed when the outer crust ruptures during oven spring; in a pan loaf of bread, it is the lighter and rougher area along the side of the loaf just above the pan top.

break-and-shred — see the second definition of "break."

breakdown — (1) The development of undesirable chemical or physical changes in a frying fat. May include darkening, formation of excess free fatty acids or peroxides, polymerization and gumming, undesirable foaming, and development of unpleasant odors and flavors. (2) The complete mechanical failure of a piece of equipment or entire line.

break flour — flour produced by the break rolls as the wheat passes through them in the milling process.

breaking down — deterioration of the physical properties of a dough, creamed mass, or batter due to excessive mixing.

break system — in a flour mill, the series of pairs of corrugated steel rollers that tear open the wheat kernels and remove endosperm from the bran.

brew — a bread dough intermediate consisting of water, yeast, yeast nutrients, a buffering agent, and often some flour, that has been fermented for a specific time at a predetermined temperature. After fermentation is complete, the brew is mixed with the remainder of the ingredients, including the flour. A brew serves many of the same purposes as a liquid sponge or a "ferment," but generally contains less flour and more water and is less viscous.

brine — usually, water saturated (or nearly so) with sodium chloride. In pickling of meats, brine may include other important ingredients such as nitrates or nitrites.

brioche — applied to a rather wide range of baked roll and loaf products; brioches are typically made from yeast dough and are rich in butter and eggs but are not particularly sweet. Often formed into large rolls or small loaves consisting of a larger bottom portion with a smaller "topknot." Sometimes, they are made with a center filling of fruit.

Brix — a scale for converting the specific gravity of a syrup into its sugar concentration. Gives only an approximation of the true sugar content if mixtures are present.

bröd — *Sweden* bread.

bromated flour — flour to which the oxidizer or maturing agent potassium bromate has been added.

bromates — the oxidizing or maturing agents potassium bromate and sodium bromate.

bromelain — a protein-digesting enzyme obtained from pineapples and sometimes used to reduce the mixing time of doughs. Has been largely or entirely replaced by fungal enzymes.

brose — a Scottish preparation made by stirring hot water, milk, or broth into, e.g., oatmeal until it forms a thick porridge. Oats in some form is usually the principal constituent, but barley or other grain meals have also been used.

broth — a fermenting liquid prepared from yeast, water, yeast nutrients, and sometimes flour. It is used to render bulk fermentation unnecessary or to greatly reduce the duration of the bulk fermentation step.

brown and serve — describes bread loaves or rolls that have been baked until the crumb "sets up," but not long enough to brown the crust; the consumer bakes the product in a home oven until optimum crust color is obtained.

brownie — a moist, chewy, dense, chocolate cookie baked in sheets and cut into rectangular pieces. Occasionally used to refer to cookies of similar physical characteristics but not containing chocolate, e.g., butterscotch brownies.

browning reaction — the so-called Maillard reaction, which involves the interaction of amino acids and proteins with reducing sugars; it produces brown colored poorly defined compounds that often have pronounced flavors. It is the principal cause of crust coloration.

brown sugar — granulated refined sugar, the particles of which have been coated with cane molasses. The refined sugar portion can be either cane or beet sugar. The brown sugar of commerce is not a partially refined sugar.

brunt farinsocker — *Sweden* brown sugar.

Brussels biscuit — a form of zwieback.

BTU (or Btu) — British thermal unit, the amount of heat required to raise the temperature of one pound of water one degree F. Equals 0.252 Cal.

buckweisen — *Ger* buckwheat.

buckwheat — the fruit, or grain, of an annual herb native to Siberia. Dark in color, roughly triangular in outline. A grain but not a cereal.

bucky — a bucky dough is tough and dense and resists extension; it tends to tear when stretched. Buckiness is characteristic of a "young" or under-fermented dough, but there are other causes, as well.

budding — one of the reproductive processes in yeast, characterized by the formation of a protuberance on the outer wall of a cell; this later expands until it is finally cut off from the original cell and begins a separate life.

budin — *Sp* a sort of cake or trifle, served like a pudding.

buffering agent — a material that, when present in an aqueous system, tends to reduce the pH change caused by addition of acids or alkalies to the system.

bulgur — also bulghur, bulgar, etc. Grains of wheat that have been soaked and heated until the starch gelatinizes, then dried. Usually cracked into relatively small granules and used in cooking somewhat like rice.

bulk fermentation — the stage in the baking process in which the dough is fermented in the condition it is removed from the mixer, i.e., before it is separated into pieces.

bulking agents — substantially inert ingredients that are added to a mixture primarily to increase its weight or volume.

GLOSSARY OF MILLING AND BAKING TERMS

bun divider — a device used mostly in retail shops for dividing a dough mass into pieces of uniform weight by first pressing the mass into a sheet of uniform thickness and then severing pieces by a number of automatically activated knives. Often combined with a rounding device.

bun pan — a sheet pan provided with many cups or shallow cavities to restrict shifting of dough pieces during movement of the pan.

buns — small (8 oz. or less) pieces of baked bread dough, sometimes in fancy shapes and/or with enriching ingredients.

buñuelos — *Sp* fried cookies in thin shapes; in one form, made from a fluid batter coated onto fanciful metal shapes and then deep-fried.

burma — *ME* see "kadaifi."

burning out — a process applied to new baking pans for the purpose of conditioning their surface so they will absorb heat better. Typically requires greasing the pan and heating it at 400°F for 30 min or more.

burr — *ME* Arabic bread made from whole wheat meal.

bursting — wild or uncontrolled breaks resulting from excessive oven spring.

butter — fatty ingredient obtained by churning sweet or sour cream from cow's milk.

buttercream frosting — rich uncooked frosting containing at least powdered sugar and butter (or other shortening) whipped to a smooth, uniformly aerated condition.

butterdejg — *Denmark* puff paste.

butterfat — the lipids that constitute approximately 80% of butter; same thing as milk fat.

butter horns — basic sweet dough cut and shaped like horns.

butter icing — a creamed mixture of butter and powdered sugar with other ingredients, forming a rich smooth oleaginous frosting or icing. Commercially, some of the butter is usually replaced with margarine or an appropriate shortening.

buttermilk — the fluid remaining after sour cream has been churned and the butter removed; this is the material that forms the basis for the commercially available dried buttermilk sometimes used as an ingredient in bakery products. Cultured buttermilk, as found in the supermarket dairy case, is generally skim milk in which bacterial cultures have been allowed to grow until the desired physical and chemical changes have occurred.

butterscotch — a candy made of brown sugar and butter cooked to a very low moisture content until the typical appealing flavor forms; also natural and artificial flavors simulating the flavors obtained in such a process.

butter sponge cake — cake made from sponge cake batter to which butter or other shortening has been added.

by-product — a material that inescapably results from the production of a desired food, e.g., whey is a by-product of cheese manufacturing.

-C-

cacahuates — *Mex* peanuts.
cacao products — chocolate, cocoa, cocoabutter, and other products made from the cocoa bean.
cadaif — see "kadaifi."
cake — a soft sweet baked product of widely variable size made from a flour-containing batter and often frosted and filled. Can be leavened with chemicals, steam, or air, or combinations of these.
cake beater — the agitator blade used primarily for mixing batters in vertical mixers. It consists of a moderately thick blade shaped to conform closely to the bowl sides and having large openings.
cake flour — a wheat flour of low protein content, having a very white color and fine particle size. These flours are nearly always treated with chlorine.
cake hoop — a circular metal ring, one to a few inches in height, set upon a baking sheet to contain cake batter during baking.
cake liners — paper or parchment cups into which cake batters are deposited prior to baking; also the discs or sheets used in the bottoms of circular or sheet pans.
cake pans — metal or oven-resistant plastic containers into which batters are scaled and carried through a baking chamber.
cake slabber — a cutter for horizontally slicing sheet cakes or the like into thinner layers.
calabaza — *Sp* pumpkin, sometimes squash.
calcium carbonate — a mineral substance used as a calcium source in yeast foods and as an enriching agent in bread, etc. An effective buffer.
calcium peroxide — a dough oxidant having somewhat different properties than the bromates or iodates.
calcium propionate — a mold inhibitor.
calcium stearoyl lactylate — a dough conditioner that will, in many cases, improve the processing response of doughs and batters and extend shelf life of the finished product.
caliente — *Sp* warm, hot.
calzone — *It* a folded pizza shell with all the "topping" inside; it can be baked or deep fried.
camotes — *Sp* sweet potatoes.
campechanas — *Mex* a pastry made of a strudel-like dough sheet which is rolled into a tube, cut crosswise, sprinkled with sugar, and baked.
canache — *Fr* a chocolate filling made of cream and chocolate liquor or other couverture, plus other ingredients. Also known as "Paris cream."
Canadian cheese — when it refers to a bakery product, this means a pastry made of cooked meringue filled with a marron-flavored butter cream.

candying — modifying a fruit piece by replacing its internal fluids with a concentrated sugar syrup. This treatment greatly changes the texture, flavor, and appearance of the fruit, but makes it resistant to microbial spoilage. In the case of candied cherries, only the shape and basic structure of the fruit remain, the color and flavor having to be replaced with additives.

canola oil — this item of commerce used to be called rapeseed oil, but a new name was chosen for public relations purposes. It is high in monounsaturated fatty acids. Ordinary rapeseed oil is high in erucic acid, which has some undesirable health implications, but the canola designation is said to be restricted to oil from rapeseed cultivars that are low in this component.

capillary melting point — the temperature at which a fat becomes completely clear and liquid inside a capillary tube of specific dimensions.

captive bakery — a bakery owned by a supermarket chain.

caramel — sugar, water, and corn syrup boiled to between about 260° to 360°F until a dark colored mass is formed.

caramel buns — sweet dough pieces (usually of the cinnamon-roll type) baked in a pan containing cups that have been coated inside with a sugar and shortening mixture.

caramel color — a very dark brown, nearly flavorless syrup made by processing corn syrup and offered commercially as an ingredient for coloring rye breads, etc.

caramelized sugar — dry sugar heated with constant stirring until it becomes dark in color; used for flavoring and coloring.

caramelo — *Sp* caramel. In the plural, small candy pieces.

caraway — crescent-shaped grayish-tan seeds harvested from a white flowered annual of the parsley family (*Carum carvi* L.). The dried seeds will average about 6 mm in length. They have a characteristic agreeable odor, and an aromatic, pleasant, warm, sharp taste. Whole and ground seeds are very frequently used in rye breads and less frequently in cookies.

carbohydrate — a chemical compound composed of carbon, hydrogen, and oxygen, generally in the approximate atomic proportions of one C, two H, and one O. Examples are starch, sugar, and cellulose.

carbon dioxide — a colorless, tasteless, edible gas that is the principle leavening agent of most bakery foods. In these products, it is derived from yeast fermentation, sodium bicarbonate, or ammonium bicarbonate, with few exceptions.

carboxymethylcellulose — a modified form of cellulose that binds large amounts of water and is sometimes used as a thickener in batters, icings, frostings, etc.

cardamon — decorticated cardamom seed is the dried, ripe fruit of *Elletaria cardamomum* L. Maton. The hard, wrinkled, light reddish-brown seed has a pleasant aromatic odor and a characteristic warm, slightly pungent taste. This spice, which is relatively expensive, is occasionally used in Danish pastries, cookies, pumpkin pies, and coffee cakes.

carob bean — also, "locust bean," "St. John's bread," etc. The sweet, succulent seedpod of the carob tree. Contains a high content of sugar. Has been roasted and otherwise treated to yield a substitute for cocoa, though it has none of the appealing chocolate aroma and contains very little fat.

carotene — a yellow pigment occurring in many vegetables and cereals.

carotenoids — natural and synthetic orange and yellow pigments of the same general chemical type as carotene.

carrageen — also, carrageenan. A water-binding and gelling agent extracted from certain kinds of seaweeds. It consists of a complex mixture of compounds, each of which has somewhat different gelling and thickening properties. Used as an ingredient in jelly-type fillings, in frosting, etc.

casein — the predominant type of protein in milk and cheese.

cashews — nuts harvested from *Anacardium occidentale*, a tree grown in the West Indies, South America, and the Far East. Most cashews entering the international market have been processed in India. A crescent-shaped seed, high in fat, mildly flavored, only moderately crisp when roasted, and light in color. Used occasionally as a topping on baked goods, but much less important in U.S.A. commerce than almonds, pecans, walnuts, and peanuts.

cassia — dried bark of the evergreen tree *Cinnamomum cassia* Blume, now used to mean practically the same thing as cinnamon, which more correctly should be restricted to material from *Cinnamomum zeylanicum* Blume (Ceylon cinnamon). Bakers usually obtain this material in ground form. It is widely used in bakery products, where its pleasant, warm, spicy odor is appreciated by almost all consumers.

castaña — *Sp* chestnut.

castera — *Jap* sponge cake.

caster sugar — *UK* also, castor sugar. An imprecise term fairly common in the UK, but seldom if ever used in the U.S.A. Essentially the same as granulated cane or beet sugar, though the particle size distribution may be somewhat different (finer in caster sugar). The name originated as a description of the kind of sugar that was placed in the shaker (caster) that stood on the dining table or kitchen table.

casting sugar — high boiled (low moisture) sugar syrups used for making molded decorative figures.

castor sugar — see "caster sugar."

catalase — a widely distributed enzyme that greatly speeds up the decomposition of hydrogen peroxide and, perhaps, of other compounds.

catalyst — a material that speeds up a chemical reaction without undergoing permanent change itself.

cebollas — *Sp* onions.

cellophane — flexible transparent film made of regenerated cellulose; has excellent optical properties, but has poor resistance to moisture transfer and poor sealability. Often combined (laminated) with other films.

cellulose — the chief structural material of all woody plants; chemically, it is a polymer of glucose like starch but with the glucose residues joined with a different bond orientation. Much more resistant to chemical attack than starch, and not digestible by humans.

cellulose acetate — a transparent thermoplastic material made by esterifying cellulose with acetic anhydride and acetic acid. It can be made in packaging films, and also be molded, extruded, and cast into various shapes.

C-enamel — a coating applied to the inside of metal cans to reduce the attack of sulfur compounds on the metal.

centeno — *Sp* rye.

centrifugal — in a flour mill, a rotating bolter in which the ground material is forced against a circumferential sieve by rotating beaters while the sieve also rotates.

centrifuge — a device that spins liquids at high speeds so as to cause separation of the liquid's components according to their density.

cereals — the seeds of grassy plants as used for food or feed; also, the plants yielding such seeds.

Cerelose — a trade name for crystalline dextrose.

certified color additives — certain chemically synthesized dyes and their lakes that have been approved by the FDA for use in foods if the manufacturing batches have been certified. There are many "uncertified" coloring agents that are also FDA-approved.

chaff — the fine, light material that separates from grain during harvesting or cleaning.

chain oven — a continuous traveling-tray oven in which the trays carrying the product are pulled forward by projections on parallel chains. The chains form an endless belt passing through the oven from front to back and then from back to front underneath the baking chamber.

chamucos — *Mex* pastry made by forming a ring of pan fino around a center plug of flavored paste, then baking.

chapati — *Ind* a type of unleavened pan-baked flat dough piece made from whole wheat meal.

chapelure — *Fr* bread crumbs.

charlotte russe — a dome shaped mold lined with ladyfingers and filled with bavarian cream.

chausson — *Fr* (1) A puff pastry dough piece garnished with apple jelly before baking. (2) A sweet or savory calzone.

chaval — (chawal) *Ind* rice.

chateau — *Fr* a wine foam served warm as a dessert or as a sauce for puddings, cake slices, etc.

cheese — the food product made by coagulating the casein of milk and then separating the curds from the by-product, whey. The curds so formed are usually pressed into a solid mass, then cured or aged over weeks or months. Among the many variations on this basic process are inoculation with bacterial or fungal cultures, adding colors, flavors, and salt, and curd washing.

cheese cake — there are many different kinds of cheese cake, but all of them contain a substantial proportion of some kind of relatively bland flavored cheese, such as bakers' cheese. Otherwise, there are hardly any unifying factors agreed upon by the experts.

chef — a culinary artist; professional cook.

chelating agent — a chemical compound that forms unusually stable complexes with metal ions.

chemical leavening — usually some form of baking powder, i.e., a mixture of sodium bicarbonate with an acid reacting substance. Ammonium bicarbonate is also in this category.

chemically leavened — a product that owes some part of its volume increase to the evolution of carbon dioxide from sodium bicarbonate or ammonium bicarbonate.

cherries — small stone fruits of certain species of trees or shrubs of the genus *Prunus*. There are sweet cherries, such as the Bing, as well as the red sour pie (RSP) cherries that are commonly used in jams, jellies, preserves, and pie fillings.

chicha — *Mex* originally an Aztec/Mayan/etc. drink made from the fermentation of corn that has been chewed and spit out. Now, applied to various kinds of alcoholic and non-alcoholic beverages.

chiffon cakes — products baked from batters resulting from combining a whipped mass of whole eggs, flour, and melted butter (or vegetable oil).

chill — the hardened external surface of a chilled-iron roller.

chlorination — (1) Adding small amounts of chlorine gas to wheat flour in order to whiten it and improve its quality for cake baking. (2) Injecting chlorine into a stream of potable water to kill bacteria and inactivate viruses; it also improves taste of the water in some cases.

chlorine — an element that exists in gaseous form at room temperature and atmospheric pressure. The gas is greenish-yellow, of relatively high density, and has a pungent, acrid, disagreeable odor. It is poisonous at low concentrations.

chocolate liquor — finely ground roasted cacao bean nibs, i.e., pure chocolate without any additives; also called bitter chocolate and baker's chocolate.

chocolate products — foods made from finely ground cacao beans mixed with other ingredients such as sugar, milk, etc. Some of the standard forms are bitter chocolate, milk chocolate, and sweet chocolate.

Choco-roles — *Mex* a trade name for chocolate-covered "Swiss rolls" containing pineapple jelly filling.

cholesterol — an important physiologically active fat-soluble compound supposedly found only in animal cells.

chop — the product of a break operation in a roller mill.

Chorleywood bread process — uses high speed development to replace the dough conditioning that normally occurs during bulk fermentation. Also relies on heavy supplementation of the dough with several chemical modifiers.

choux paste — a type of dough or batter used for making eclair and cream puff cases. Produced by beating eggs with a mixture of fat and gelatinized (with hot water) flour. When properly baked, large bubbles form inside the piece and then collapse so as to give an almost empty interior.

chrust chyli faworki — *Poland* "kindling or favors," a kind of fried cookie typically made of flour, butter, sugar, egg, vinegar, and sour cream; not leavened; an Easter dessert.

chunk — a grain particle composed mainly of bran, but having some endosperm attached. These particles are produced when milling conditions are too extreme.

churro — *Sp* a more or less sweet fried dough usually in stick form (often with longitudinal corrugations) but sometimes in doughnut shape.

cinnamon — a flavoring material made from the bark of the evergreen tree, *Cinnamomum zeylanicum* Blume. Ceylon cinnamon. This name is often applied to cassia, and both products are used in the same way.

C.I.P. systems — clean-in-place systems include vats in which cleaning solutions are prepared, high pressure pumps for transferring the solutions, and spray balls inside the food processing unit to disperse the solution to all surfaces requiring treatment. Draining and collecting means are also required for the effluent.

ciruela — *Sp* plum.

citric acid — an organic acid found in many fruits, such as oranges and lemons. Most commercial citric acid has been prepared by oxidation of glucose. Large quantities of this material are used in the beverage industry, but it is less important in bakery formulations.

citron — a fruit of the citrus family now used only as a source of rind for candying.

clarify — (1) To prepare drawn butter by melting butter, allowing the sediment and water to collect at the bottom, then pouring off the clear supernatant, which consists almost entirely of butterfat. (2) To remove insoluble materials from fruit juices and the like by filtration or settling.

clé — *Fr* the outer seam in a dough strip that has been molded into a loaf.

clear flour — the portion of flour remaining after the "patent" mill streams have been diverted.

clearing time — time from beginning of mixing until the dough forms into a single mass and takes up the material smeared on the back of the bowl.

clears — the coarser parts of a straight flour. These mill streams are sometimes divided into first clears, second clears, etc.

close-textured — describes the interior of a loaf or roll that has uniform and small vesicles.

clove — a spice of penetrating aroma and hot pungent taste, made by drying the unopened flower buds of a tropical tree, *Eugenia caryophyllata* Thunb. It is used fairly widely in bakery products, probably mostly in mince pie fillings.

cobbler — a baked dessert something like a pie, but generally baked in a deeper pan and sometimes having only the top crust. May have the top crust formed of a sweetened soda biscuit dough.

cocoa — a powder made from chocolate from which most of the fat has been pressed or extracted.

cocoabutter — fat pressed from ground (and usually roasted) cacao nibs.

coconut — commercially, dried shreds or granules made from the meat of the fruit of the coconut palm.

coconut oil — the fat pressed from the dried meat of the coconut. It is a bland white fat of low melting point that is very useful in icings, etc. It is highly saturated.

Codex Alimentarius — a collection of international ingredient standards or specifications.

coffee cake — a multi-serving size of baked sweet yeast-leavened dough made in various shapes and, usually, with fillings or toppings. Dough formulas may vary from fairly lean to very rich. Both chemically leavened and fermented dough varieties are known.

colchones — *Mex* orange-flavored sweet bread.

cold test — a test that determines the extent to which the high melting-point fractions of an oil have been removed during the production of a "winterized" oil. In this test, the oil is held in an ice water bath and the time required for the first appearance of cloudiness is recorded as "Cold Test Hours."

coliforms — gram negative bacteria that can ferment lactose. The most familiar member of this large group is *Escherichia coli*, found in vast quantities in the gut of man and many other animals.

combination bakery — a bakery that uses more than one form of production or supply, e.g., a bakery that uses both refrigerated dough bake-off and scratch mix production methods.

compact oven — a smaller version of the revolving tray oven in which the trays revolve from side to side, instead of being carried from front to back. This allows placement of the oven in a shallow space that is more convenient for many food service operators and bake-off boutiques.

composite flour — wheat flour blended with some nonwheat meal or flour (such as rye flour).

compound — (1) A chemically defined substance consisting of two or more elements combined in fixed proportions. (2) A prepared mixture of some kind.

compound coatings — imitation chocolates made by combining (for example) cocoa, coconut oil, and sugar. Other flavors besides chocolate can also be made in this way. The essential difference between true chocolates and compound coatings is that little or no cocoa butter is included in the formulas of the latter.

compound shortening — a mixture of animal and vegetable fats processed as shortening.

compressed yeast — consists of undried cells of living yeast that have been combined with fillers and pressed into cakes.

compression board — in a molder, a strip or plate that applies pressure to the curled piece of dough as it is carried beneath the strip or plate; can be adjusted to vary the force applied to the dough piece.

compression chamber — (1) In a dough divider, the cavity of defined dimensions into which dough is forced to measure the desired portion size. (2) In a dough molder, the section where pressure is applied to the curled piece of dough so as to seal the layers together.

compressor — the pump in a refrigerating unit; it raises the pressure of the refrigerant before it flows to the condenser.

conchas — *Mex* rounded pieces of pan fino made up in the shape of a sea shell and sometimes topped with sweet paste.

condensation — in general, the transition of a substance from the vapor phase to a liquid state. Also, the formation of water droplets ("dew") on the surface of a relatively cold solid surface that has been exposed to a warmer atmosphere.

condensed milk — whole fluid milk from which a substantial portion of the water content has been removed by evaporation. Usually contains a high percentage of added sugar.

condenser — the unit in a refrigeration system that receives the hot, high pressure refrigerant gas from the compressor and cools it until it returns to the liquid state.

condiment — a material that can be added to a prepared food to give it added zest or appeal; mustard, ketchup, and pepper are examples.

conditioning — controlled moistening of grain to prepare it for grinding; the bran is toughened and the endosperm softened by the added moisture. Heating is involved in some of these operations. Similar to tempering.

conduction — a method of heating or cooling in which the basic physical mechanism is the transfer of energy from one solid or liquid material to a contacting solid or liquid material.

confectioners sugar — granulated sugar that has been finely ground to a smooth powder and sifted; available in two or more particle size distributions; normally contains about 3% cornstarch to prevent caking. Synonymous with "powdered sugar" and "icing sugar."

confectionery fats — a broad range of ingredients used in the formation of sweet confections, including some bakery products. Their primary application is in the formulation of compound coatings, both chocolate-substitutes and white and colored coatings.

confectionery — prepared sweet dessert-type foods including all kinds of candy and some bakery products.

congeal — to change from a liquid into a solid (or semi-solid) mass.

consistency — as applied to dough, means the tactile evaluation or "feel' of the dough. It is one subjective criterion by which the proper absorption is judged.

contre-frase — *Fr* making the dough tougher by adding flour to it during kneading.

control — a sample or experimental specimen that represents the standard material before any of the variations being tested have been applied.

controlled atmosphere packaging — a method of extending the shelf-life of foods by reducing or eliminating the amount of oxygen in the package and, sometimes, by adding gases such as carbon dioxide.

convection — heat transmission by moving currents of gases or liquids.

convection ovens — any oven in which the main heat transfer method is moving air. The addition of fans and blowers provide forced convection that speeds up baking.

conveyor — any type of fixed-in-place device that can continuously transfer material from one location to another; rollers, belts, augers, buckets on chains, and air flowing through pipes are some of the common moving forces and mechanisms.

cookie — a sweet baked product typically containing flour, sugar, shortening, flavoring, and other ingredients, in small piece size, and relatively low in moisture content. Generally, cookies are relatively dense as compared to cakes. Characteristics of examples of this class vary enormously.

cookie bag — a canvas or plastic bag of roughly conical form that can be filled with dough and squeezed by hand to force the dough through a metal or plastic orifice at the tip of the cone.

cookie cutter — any type of hand-operated device that can be used to cut shaped pieces from a sheet of cookie dough.

cookie sheet — a flat metal sheet, usually without elevated rims, that can be used to support cookie pieces while they are being transferred into the oven, and to hold them during baking.

cooler — (1) A refrigerated chamber held above freezing temperatures. (2) A conveyor or chamber used for bringing hot loaves and rolls to about room

temperature, usually by exposing them to currents of ambient air.
cooling conveyor — usually a system of belts suspended from the ceiling and equipped with a fan system that draws in air at the discharge end and moves it over hot loaves that have just emerged from the oven.
core — in describing baked product texture, this refers to condensed or solid regions within the crumb that appear to have not undergone any significant expansion. A serious quality defect.
coriander — a spice made from the dried fruit of *Coriander sativum* L. The globular, yellowish-brown seeds have a slight fragrant odor and a pleasant taste. Although it has been used in many kinds of bakery products, it is not one of the favorite spices for this purpose in the U.S.
corn — the plant *Zea mays* and its seed. Exists in various types such as sweet corn, field corn, and popcorn. In the U.K., the term corn may be applied to almost any kind of grain.
corn bran — the fibrous outer coating of the corn kernel, regarded as a low-grade food for cattle or a high-grade food for humans.
corn cones — ground corn of particle size intermediate between meal and flour, used in the bakery for dusting dough pieces and pans.
cornet — *Fr* a cone-shaped pastry.
corn meal — meal of various granulations produced by dry-milling kernels of white or yellow field corn, usually with the germ removed to delay development of rancidity.
corn meal boards — plastic, fiberglass, or wooden sheets used to support molded dough pieces during transfer and proofing, so called because cornmeal is dusted on the surface of the sheet to prevent adhesion of the dough. Also called proofing boards.
corn muffins — slightly sweet muffins containing varying proportions of corn meal and wheat flour. Chemically leavened.
corn oil — the edible oil pressed from corn germs.
corn starch — a fine white powder made from corn kernels by the wet milling process.
corn sugar — dried glucose made from corn syrups.
corn syrup — viscous liquids containing mixtures of sugars made by the controlled hydrolysis of corn starch; many different kinds are available — they vary in the kind and amounts of carbohydrates they contain, the total concentration of solids, etc.
corn syrup solids — corn syrup that has been dried and made into a fine powder.
cottonseed flour — milled cottonseed cake that results from the oil pressing operation. Usually heat treated to deactivate certain toxic factors (gossypol) in the meal.
cottonseed oil — the edible refined oil pressed or extracted from cottonseed. Highly regarded as a constituent of hydrogenated shortenings.

coup — *Fr* dish containing a single scoop of ice cream which is usually decorated with syrups, fruit pieces, and whipped cream.

coupe-pâte — *Fr* bowl scraper or dough knife.

couper le pâton — *Fr* to slit the dough just before it enters the oven.

couronne — *Fr* crown-shaped loaf, formed by baking the dough in a basket having a center tube.

couverture — enrobing materials of which chocolate is the prototype; pastel-colored fatty coatings of non-cacao origin are also placed in this category.

cracknel — a plain cracker made originally of only white flour, eggs, and sugar (no water); various shapes. The best examples have an unusually low density and a very fine grain, and exhibit a smooth and shiny surface.

crack stage — in sugar boiling, the condition reached by a syrup brought to 280°F. If a drop or string of the sugar is cooled in cold water, it cracks when deformed.

cream — (1) A fluid milk product enriched in fat to different levels, such as half-and-half and whipping cream. (2) A thickened cooked mass of (usually) sugar, egg, milk, and a viscosity improver (starch, gelatin, etc.) used for filling pies, doughnuts, etc. A flowable pudding.

cream horn — a hollow cylinder formed from a spirally wound strip of puff pastry and filled (after baking) with whipped cream or the like.

creaming — the process of mixing and aerating a shortening and one or more dry ingredients, such as sugar or flour.

cream injector — a device for extruding cream filling from a reservoir through a tube into the centers of baked or fried dough pieces such as doughnuts and snack cakes.

cream of tartar — acid potassium tartrate; one of the acid-reacting substances used in baking powders. In the ingredient form, it is a white powder. Also used in small amounts to facilitate the whipping of egg whites and as a "doctor" or sugar hydrolyzing agent in the boiling of sucrose syrups.

cream pies — pudding or custard type fillings poured into pre-baked pie crusts; usually topped with whipped cream.

cream puffs — hollow balls of baked cream puff dough (choux paste) that have been filled with starch-based pudding, cooked custard, or whipped cream.

crema — *Sp* cream.

crème chantilly — dairy cream whipped with vanilla and sugar; used as a filling or topping for sweet baked products.

crepes — thin pancakes of the French type.

crescent roll — a yeast-leavened bread roll made by rolling up a triangle of sheeted roll-in dough so that the center is much thicker than the ends, then curving the dough piece into a crescent shape.

cripple — a damaged or otherwise defective and unusable baked product.

croissant — now often used as a synonym for crescent roll, but formerly used only for similarly shaped pieces made with a rich but not sweet roll-in dough. Yeast-leavened.

croquembouché — small cream puffs stuck together in the form of a pyramid or cone, using caramelized sugar as the adhesive. Often, elaborately decorated.

cross-grain molder — a loaf molder that takes the sheeted dough and transfers it to a forming conveyor at right angles to the flow of the sheeting conveyor. Its purpose is to roll up the dough perpendicularly to the direction of sheeting, so as to make the moisture distribution in the finished dough piece more nearly uniform.

croutons — in present day commercial bakery usage, "croutons" is generally understood to mean cubed dried bread that has been seasoned with herbs and spices and sprayed with oil to serve as an additive to soups or as a basis for poultry stuffing.

crown — inside top of the baking chamber of an oven, especially applied to large stationary hearth ovens.

crullers — fried dough products usually in the form of long (6 to 8 in) double twists, though sometimes in a fancy doughnut (circular) shape.

crumb — in leavened baked goods with a crust, all of the dough product except the crust.

crumbs — small fragments formed in any manner from larger pieces of baked products.

crumpet — (UK) the distinction between crumpets and English muffins is unclear. Both are made from very thin yeast-leavened doughs or batters, and both are generally cooked on a griddle, often in restraining rings. Both are split and toasted immediately before consumption. As sold in mass markets, most crumpets appear to be moister and more gelatinous in texture than muffins and to have many more large holes (bubbles) in their interior; their crust is not as clearly differentiated from the crumb. Some crumpet recipes call for eggs, never seen in (English) muffin recipes.

crust — the distinctive outside layer of a baked product. Includes all of the portions that have undergone browning or substantial dehydration and consolidation during baking. By extension, a hard or crisp layer on any normally soft piece of food.

crusting — the formation of dry surface layers on dough pieces due either to loss of moisture or accumulation of dusting flour. Generally, very undesirable.

cryogenic freezer — freezer that uses liquefied carbon dioxide or nitrogen to withdraw heat from products.

crystallized — said of aqueous sugar solutions that have deposited crystals from a supersaturated condition; also describes fruits and the like that have been steeped in saturated sugar solutions and, finally, rolled in or sprinkled with sugar crystals; used for decorating and otherwise enhancing bakery foods and confections.

crystallization — condensation from solutions or molten masses of particles which exhibit uniform arrangements on the atomic or molecular scale and that are nearly pure.

cube sugar — pure cane or beet sugar that has been crystallized in blocks or sheets and then cut into cubes. Other methods include crystallizing the sugar directly into cube molds.

cuernitos — *Mex* relatively small cuernos.

cuernos — *Mex* crescent shapes of pan fino, with flavor paste applied to the outside before baking.

cup — a standard measuring cup, as used in many consumer recipes, is 8 fluid ounces. The volumes contained by other kinds of cups cover a wide range.

cup cakes — small sweet muffins baked from cake-type batter deposited in muffin pans, often iced and decorated, sometimes filled.

curacao — an alcoholic liqueur that has been flavored with dried orange peel; occasionally used for flavoring gourmet icings, fillings, etc.

curd — the casein lumps that form when milk is coagulated with rennet and/or acids.

curdled — batters that appear to have separated into liquid and pasty fractions, somewhat resembling curdled milk.

curling chain — in a bread molder, a chain belt that pulls up the leading edge of a dough sheet and begins the curling action that eventually forms a cylinder.

currants — the acidulous berry of the currant bush, somewhat similar to gooseberries, but in the baking trade "currants" means small raisins.

custard — basically, a sweetened mixture of eggs and milk that has been cooked over hot water. Hundreds of recipes exist.

cut-in — to mix solid fat and flour with a pastry blender, fingertips, or machine so that the shortening lumps are flour-covered but separate.

cut-off — adjustable dividing board located under a sieve to enable the miller to change the flow of stock; purpose: to get the best possible results as grinding and sifting conditions change.

cutting roller — the more rapidly rotating member of a pair of cylinders in a flour mill roll-stand.

cysteine — a sulfur-containing amino acid. It is used as a reducing agent to modify the physical properties of doughs.

-D-

dalchini — *Ind* cinnamon.

damasco — *Mex* apricot.

damper — an adjustable plate or other device to control the flow of smoke or air to or from a combustion chamber.

Danish pastry — originally, a very rich, flaky yeast dough that had been layered with butter or other shortening. Now, often refers to any kind of sweet breakfast pastry, even those not made from roll-in dough.

dark meal — in cookie bakery parlance, means the ground-up scrap of dark-colored and strong-flavored cookies that have been rejected (or returned) for some reason. Used as a bulking or non-characterizing ingredient in dark-colored doughs such as molasses cookie formulas.

date filling — ground or chopped dates mixed with water and other ingredients and cooked.

DATEM — diacetyl tartaric acid esters of monoglycerides; a class of chemical compounds that affect the properties of dough; they are employed to increase loaf volume and soften the crumb.

dater — an apparatus used to imprint a date code on a package.

dates — the fruit of a species of palm; there are several varieties, all very sweet and with low acidity.

deck oven — small oven with stationary hearth, often heated by electricity. Loaded and unloaded by peels. Widely used in pizza shops.

decorating tubes — metal or plastic orifices to be slipped into the opening at the point of pastry bags for the purpose of forming the extruded icing into fancy shapes.

defatted soy flour — soybean meal from which substantially all the fat has been removed; it is almost always heat-treated to inactivate undesirable components and improve flavor. Used as an inexpensive protein supplement and to bind water in doughs and batters.

defect — failure of some quality factor of a product (or ingredient or package) to reach an acceptable level.

degasser — a machine or device that works the dough prior to dividing so that much of the gas can escape; usually it is a kind of pump or kneader placed at or near the divider hopper.

degermed — having the plant embryo removed; a term applied almost exclusively to corn and rice.

degree — a unit on a scale of measurement such as Fahrenheit for temperature, Brix for sugar concentration, etc.

dehydrate — to remove water from a substance by any method, although not usually applied to procedures that strain or press liquid water from a mass.

delidder — a machine for automatically removing the covers from pans of sandwich bread and the like.

demi-glacè — *Fr* soft frozen ice cream.

dendritic salt — salt that has been crystallized with additives so that it forms branched crystals.

density — mass per unit volume.

deodorization — the last step in traditional processing of fats and oils. It removes the relatively volatile trace components (such as ketones, aldehydes, alcohols, and free fatty acids) that contribute color, odor, and flavor. Typically, deodorization gives an oil that has less than 0.05% free fatty acid content and that is nearly flavorless and odorless.

depanners — machines that remove baked product from pans and, usually, convey them to packaging equipment.

deposit cookies — cookies made from very soft doughs that are extruded directly onto the oven band and, usually, not cut off by a knife or wire.

depositer — a scaling device that extrudes or drops a measured amount of material (such as muffin batter) into a pan or other receptacle.

derivative — a substance obtained by chemically combining one compound with another.

desiccate — to dry by removal of water vapor.

detergent — a surface active agent, other than soap, that can be mixed with water to increase the effectiveness of the fluid in removing grease, soil, and dirt from a surface.

developing — the process of mixing a bread dough (or the like) for a time, at the speed, and under the conditions required to cause the dough to exhibit the best processing response of which it is capable. If the flour is of good quality and the formula balanced, a bread dough can be developed to a soft but very elastic and extensible mass that retains leavening gases and goes through the make-up machinery without becoming weak and sticky. Developing involves both chemical and physical changes, some of which are poorly understood.

devil's food cake — a chocolate cake that has the leavening system adjusted so that the batter and cake crumb has a markedly alkaline reaction, leading to the development of a so-called mahogany color of the crumb.

dew point — the temperature at which moisture from the air will condense on a surface; an indication of the relative humidity of the air.

dextrin — an industrially useful material made by treating starch with acid or enzymes to partially hydrolyze it; has a low flavor (practically no sweetness) but is useful as a bulking agent, adhesive, and viscosity adjusting ingredient.

dextrose — another name for commercial grade D-glucose, or corn sugar.

dextrose equivalent (DE) — a measure of the content of reducing sugar in a syrup or powder; it is calculated as though all of it were dextrose and reported as a percentage of the total dry substance in the ingredient; one of the essential specifications for corn syrup.

dhania — *Ind* coriander.

diacetyl — a compound with a buttery flavor, present in doughs as well as in butter. It can be procured from flavor supply houses as a partially purified substance.

diastase — an enzyme that can convert starches into dextrose and maltose; synonymous with "amylase," which is the preferred term.

diastatic malt — malt or malt syrup that retains a considerable amount of the original amylolytic power of the sprouted barley from which the ingredient has been prepared; the more severe the heat treatments applied during manufacture, the lower the diastatic power of the malt.

diastatic power — a numerical designation of the starch-hydrolyzing power of an amylase preparation, such as malt syrup, as determined by a standardized test.

dielectric heating — a method of baking or cooking by a high frequency electromagnetic field that generates heat by causing rapid movement of some of the molecules making up the product. Has been commercially used to further dehydrate cookies or snacks that have passed through an ordinary oven.

dies — shaped metal or plastic orifices or cavities that are used in cookie manufacure to give the desired form to doughs.

dietary fiber — a food constituent that passes through the human intestinal tract without being digested. Includes not only true fibers such as cellulose, but materials such as pectin.

dietetic — identifies a food that allegedly has some nutritional modification of interest to health-minded consumers. Not to be confused with "diabetic" or "reduced calorie," which are terms of much narrower significance.

differential — the ratio of the speed of rotation of fast and slow rolls in a pair of cylinders.

differential scanning calorimetry — a technique for the study of heat flow in substances undergoing changes of temperature or phase. Such heat flow may result from changes in physical state (as from a liquid to a solid) or from polymorphic crystal transformation.

diglyceride — a chemical combination of fatty acids and glycerol in the proportion of two fatty acids to one glycerol residue.

dimethylpolysiloxane — an antifoam agent used to reduce processing problems that arise when cooking products that have a tendency to foam excessively.

dinner rolls — bread rolls of almost any kind of shape and size.

direct fired ovens — baking equipment in which combustion occurs within the oven chamber; allows smoke and vapor from the fire to come in contact with the dough, which is generally undesirable.

disaccharide — a sugar, such as sucrose or maltose, made up of two monosaccharides, such as glucose or fructose, joined by chemical bonds.

disc separator — a machine containing a set of upright revolving discs covered with small pockets, used in grain cleaning to remove foreign matter that has either a size or shape different from that of the desired kernels.

disinfectant — a chemical that can be applied to equipment or other surfaces to reduce the number of microorganisms present.

disperse — to distribute particles of a substance more or less uniformly throughout the total mass of another substance, as when spices are dispersed in a premix or nut granules are dispersed in an icing.

dissolve — to cause a solid material to separate into its constituent molecules (or ions) and distribute uniformly through a liquid, as when sugar or salt is dissolved in water.

distillation — converting a liquid (often present as part of a complex mixture) into vapor, and then condensing the vapor and collecting the relatively pure liquid.

disulfide bond — when two sulfhydryl groups (-SH) on proteins are close together, they may, under certain circumstances, react to form a disulfide bond (-S-S-) that joins two molecules or two parts of one protein molecule.

divider — machine that cuts masses of dough into pieces of uniform weight.

divider oil — oil used to lubricate the parts of a divider that contact dough; usually a highly refined mineral oil, although some vegetable oil compounds are also being used.

dobosh torte — or, "dobos torta." A confection consisting of several thin layers of cake alternating with layers of icing, the whole coated with chocolate. Originally, this torte was coated with brown caramel, but this treatment is seldom seen nowadays.

docker — any type of utensil or machine that has the function of punching holes from the top to the bottom of dough pieces.

docking — punching a number of vertical small holes in a dough piece with different kinds of implements, the purpose being to restrain puffing or expansion in the oven so as to get a more uniform thickness and surface appearance in the baked piece. Soda crackers are good examples of the effects of docking.

doka — *Egypt* an oiled griddle used for baking fiti, senesen, etc.

Do-Maker — a trade name for the continuous bread making-plant developed by Wallace and Tiernan.

dora-yaki — *Jap* bean jam pancakes, a sweet bun popular in Japan.

dosai — *Ind* slightly crisp, round pancake made from a fermented mixture of cereals (usually rice) and black gram.

dot — to place small bits of butter, fat, cheese, or the like over the top of a sheet of dough.

GLOSSARY OF MILLING AND BAKING TERMS

double-lap traveling oven — a large version of the traveling tray oven, in which the trays are drawn along a track that carries them both vertically and horizontally, so they can make more than two trips through the chamber before being unloaded.

double-panned — placing a cake pan containing batter on top of an inverted (empty, of course) cake pan, so as to reduce the contribution of bottom heat to the baking process.

dough — the term always applied to the extensible plastic mass formed from bread ingredients, but also applied sometimes to plastic cookie masses, pie crust mixtures, and the like.

dough brake — heavily built machines for pressing a sheet or large chunk of dough between metal rollers that rotate fairly rapidly; their function is either to squeeze out most of the gas and orient the fibrils of the gluten as part of the processing of bread doughs and the like, or to sheet out dough and fat laminations for puff pastry and the like. "Brake" is the correct spelling, but "break" is seen so often that it probably should be regarded as an acceptable alternative spelling.

dough chute — an opening in the floor of the mixing room through which a mass of dough can be dropped to the vicinity of the depositor or divider on the lower floor.

dough conditioner — a substance or mixture that improves the physical and/or processing quality of a dough when a small amount of it is added to the dough.

dough hook — for a vertical dough mixer, an agitator that is a single heavy metal rod shaped to conform to the side and the bottom of the mixer bowl.

doughnut glaze — a mixture of (usually) sugar, water, stabilizer, flavor, and color applied to the surface of cooked doughnuts by dipping or pouring. Not necessarily transparent.

doughnuts — covers a wide range of fried sweet dough (and batter) products, chemically leavened, yeast-leavened, or (in a very few cases) air-leavened. The cooked products can be glazed, frosted, iced, dusted, covered with dragees or nut pieces, filled, and otherwise modified.

doughnut screens — screens with lifting handles on each side; they are inserted into the deep fat fryer so that the doughnuts can be removed all at one time when they have been sufficiently cooked.

doughnut sticks — wooden dowels about a foot long and perhaps 0.5 inch in diameter, used for turning doughnuts in the cooker after one side has been fried.

doughnut sugar — a blend of powdered sugar, oil, and starch, often with some flavor such as vanillin, that is used to coat cake doughnuts.

dough press — a dividing machine for buns, suitable for retail bakeries, which presses a measured quantity of dough into a sheet of relatively uniform thickness and then pushes knives (dies) through the sheet to cut bun pieces of the proper weight. Also, sometimes used to refer to the quantity of dough processed in each operation.

dough room record — a paper form on which are entered the data generated by personnel (the mixer, scaler, etc.), as they perform their duties in an around the mixing machine.

dough scraper — a dull metal blade perhaps 6 inches long, with a wooden handle attached to one of the long sides, used by bakers to scrape the bench clean of adhering dough and to cut dough by hand.

dough sheeter — a machine consisting of two powered horizontal steel rollers, means for adjusting the space between them, and input and output conveyor belts. Used to reduce the thickness of dough sheets and to convert a mass of dough into sheeted form.

dough splitter — a device for cutting a slit across the top of loaves or rolls before they enter the oven; can be based on metal knives or high pressure water jets.

doughy — soft, sticky, elastic texture reminiscent of unbaked dough.

dragées — small candy pieces pan-coated with sugar, mildly flavored but often highly colored or coated with gold or silver foil. Used for decoration.

drawplate ovens — ovens in which the entire hearth rolls out the front of the baking chamber for loading and unloading of loaves.

dress — to sift or bolt.

dressing — (1) Scalping off oversize particles from a flour stock. (2) A combination of stale bread pieces, spices, and broth, baked and used as an accompaniment to meats (especially roasted poultry).

drop cookies — cookies made by dropping portions of dough on to the baking pan or band; can be done by hand or by a cookie depositor.

dropping point — the temperature at which a sample of fat becomes sufficiently fluid to flow under the conditions of the test. A portion of molten fat is introduced into a sample cup, crystallized, and then heated at a constant rate. The temperature at which the sample is able to flow through an orifice in the bottom of the cup is the endpoint.

drum molder — a machine that forms bread loaves (or other kinds of bakery products) by pressing dough pieces between a rotating drum and an outer jacket positioned to leave a channel between itself and the drum.

dry mix — a mixture of all (or most) of the ingredients, except water, required to make a particular bakery product. In some cases, the mix will contain only the basic ingredients such as flour, leavener, salt, sugar, etc., allowing the baker to add modifying ingredients to make many kinds of products from one basic mixture.

dry pack apples — peeled and cored apples canned without added water.

dry steam — steam containing no droplets of water.

dry weight basis — the expression of the weight of an ingredient or product after its content of moisture has been deducted.

duchesse — a European confection made of filberts, sugar, and egg whites and filled with nougat.

dulce — *Sp* candy; sweet.

dull-to-dull — term applied to the operation of roller mills when they are run so that the longer face of the corrugated cutting edge on the faster rotating roller meets the shorter face of the edge on the slower roller.

dummy — metal, wood, cardboard, or plastic structure shaped like a cake and used as the basis for applying decorations so the finished product can be used as a display.

dumping — (1) Pouring ingredients into the mixing bowl. (2) Decanting dough from the mixer into a trough. (3) Depanning baked cakes or bread.

dumplings — (1) Lumps of dough, not necessarily wheat flour dough and not necessarily leavened, designed to be cooked in a liquid (such as soup); sometimes, strips cut from a rolled out dough, like a giant noodle, and used in the same way. (2) As in apple dumplings, a single pared apple, usually seasoned with cinnamon, etc., wrapped in pie crust dough and baked.

dunst — middlings of small particle size from which all the bran has not been removed. Ordinarily indicates middlings that will pass through a No. 8 or No. 9 silk, but will be retained on a No. 10 or No. 11. (Obs. English).

durazno — *Mex* peach.

durum — a kind of wheat (*Triticum durum*) grown specifically for use in making a semolina that is particularly suitable for pasta products but not very satisfactory for making bakery products.

dust collector — various types of equipment for the removal of flour and other dust from the atmosphere; one type is based on bags or other types of filters through which the air is passed, another type is the cyclone separator.

dusting — (1) Separating flour from middlings, especially so that the latter may be more efficiently purified. (2) Applying flour, starch, corn meal, etc., to dough pieces and to baking surfaces to reduce sticking.

dusting flour — often a low grade flour that is procured specifically for dusting purposes.

dutch process cocoa — cocoa that has undergone an alkali treatment at some stage during its preparation — usually the cocoa beans are mixed with alkali before they are roasted. Dutched cocoa is darker and more readily dispersible in water than is "natural" cocoa powder.

-E-

éclairs — small oblong cakes, often made of cream puff dough, filled with either whipped cream or pastry cream and often iced with chocolate. Also, now often used for fancy iced and filled doughnuts of various shapes.

eesti suhrukook — *Estonia* a rich sugar cake, chemically leavened.

egg wash — lightly beaten eggs, or a mixture of whole eggs and milk, which is brushed on the outside of proofed dough products just before baking so that the finished food will have a glossy brown crust.

eier — *Ger* eggs.

eierkückas — a pancake made of batter enriched with fresh cream, a specialty of Alsace.

eish shami — *ME* a product resembling pita bread.

eigelb — *Ger* egg yolk.

eiweiss — *Ger* egg whites.

elasticity — tendency of a material to recover its original shape after it is released from a deforming force. In dough, this property is influenced by absorption, mechanical development, fermentation, and other factors. Batters normally are not elastic.

elote — *Sp* fresh (or sweet) corn.

embossing machine — device that applies a pattern to the top of a sheet of cookie dough; consists of a rotating drum with carved impresssions or molds on the surface, or sometimes patterns formed from metal strips.

emergency dough — a yeast-raised dough prepared according to a formula that allows short-cuts in fermentation and conditioning so the batch can be finished in a relatively short time.

empanada — *Mex* a kind of fried pie with a non-flaky dough folded over a filling such as pumpkin; generally small in size.

emulsifier — (1) An ingredient that enhances the formation of relatively stable systems consisting of finely dispersed globules of fat in aqueous solutions, or vice versa. Sometimes used as a synonym for surfactants, which usage, however, lacks precision. (2) A machine that forms emulsions from fatty substances and water, either by very high speed mixing action or by impinging the ingredient mixture on a plate (homogenizer).

emulsions — systems of small droplets distributed throughout a continuous phase of an essential immiscible liquid. Very common in foodstuffs; mayonnaise is a typical example. It is not strictly correct to apply the term to structures consisting of gas bubbles in a continuous liquid or solid phase, although this is often done.

endosperm — the starchy white interior material of grain; this is the material the miller desires to separate and purify to yield the finest flour.

English muffins — a yeast-raised, lean-dough bread product, in circular form, about 3 or 4 inches in diameter, with a flat top. The dough is very soft and is baked on a griddle or band, the spread usually being constrained by metal circles.

enriched bread — bread made with enriched flour and containing federally prescribed amounts of certain vitamins and minerals.

enriched flour — wheat flour that has been supplemented with certain vitamins and minerals so that it meets FDA specifications for these factors.

enrichment concentrate — the vitamin and mineral mixture that is added to bread dough to meet the requirements for enriched bread; usually sold in the form of a powder or tablets.

enrober — a machine that coats the surface of a product, such as a cookie, with melted chocolate or some other fluid material that sets up when cooled or dried.

enzyme — a class of proteins found in all living things that speeds up chemical reactions, i.e., an organic catalyst. There are thousands of different kinds. e.g., amylases which act to break down starch and proteases which stimulate the hydrolysis of proteins.

epi — *Fr* long loaf that has been cut from the edge toward the center several time along both sides, a pattern said to resemble heads of grain on a stalk.

eriste — *Turkey* noodles.

essence — a flavoring ingredient, often a concentrated extract of a natural product.

essential oil — the type of flavoring material found in many spices and extractable by ether and similar solvents; these oils are not triglycerides.

ester — an alcohol to which one or more fatty acids have been chemically bound. The most commonly found ester is a triglyceride, where an glycerol (the alcohol type substance) has three fatty acids attached. Natural and artifical flavors often contain esters.

ethanol — the alcohol formed by yeast during fermentation; it is also called grain alcohol; the same compound is the intoxicating component of beer, wine, whiskey, etc.

ethoxylated monoglycerides — a type of emulsifier made by reacting ethylene oxide with the free hydroxyl groups on monoglyceride molecules. Improves loaf volume but is not very effective in softening the crumb or increasing shelf-life.

ethyl vanillin — a synthetic flavor often used in imitation vanilla flavors; it is very aromatic but has a somewhat different character than vanillin.

eutectic — a mixture of two or more substances in a ratio such that the mixture's melting point is the lowest possible for a combination of those substances.

evaporated apples — peeled, cored, and sliced apples that have been dried to a moisture content of about 24%; usually treated with sulfur dioxide to prevent browning.

evaporated milk — unsweetened, heat concentrated milk in cans. Not much used as a bakery ingredient at this time. Has been largely replaced by dried milk.

evaporator — (1) That part of a refrigeration unit in which the refrigerant fluid is allowed to vaporize and absorb heat. (2) Any equipment designed for evaporating liquid.

extensibility — the extent to which a material may be deformed without rupture by pulling or by some other process involving the application of tension. A high degree of extensibility is the result of low yield value and high mobility associated with adequate cohesion.

extract — a flavoring ingredient, ostensibly prepared by steeping a natural material, such as vanilla beans, in alcohol or other solvent, then removing all the insolubles; that which remains is the extract. Has been applied, incorrectly, to dissolved mixtures of chemicals and natural materials.

extraction — in milling, the percentage of flour or meal obtained, using the weight of the mill mix as 100%. An extraction of 70% means that 70 lbs. of flour have been obtained for each 100 lbs. of wheat going into the mill stream; the percentages are usually adjusted to the same moisture content.

extrude — to force a plastic material (dough, for example) through a restricted opening.

-F-

facultative anaerobes — microorganisms that can grow and reproduce in the presence or absence of oxygen.

fairy cakes — *UK* iced or frosted small rich cakes made of chemically leavened and beaten egg-containing batter. Similar to petit fours.

Falling Number Test — a test used to measure the level of diastatic activity of a flour; it has some value for determining the suitability of flour for breadmaking.

family flour — a flour designed to be used in many different kinds of products, breads, rolls, biscuits, cakes, doughnuts, etc. Not the flour of choice for a professional baker, since it does many things at a barely acceptable level, but is optimum for nothing. It is usually made from hard red winter wheat of low protein content, but in some cases from soft wheat.

fancy patent — a very short patent flour, representing say 60% of the total flour produced, the rest being clears, etc.

farina — coarse endosperm particles (middlings) from the wheat milling process. Used as the basis of a hot porridge (Cream of Wheat) and in a few bakery products.

farina di avena — *It* oatmeal.

farina di frumento — *It* wheat flour.

farine — *Fr* flour.

farine d'avoine — *Fr* oatmeal.

farine de froment — *Fr* wheat flour.

farine supérieure — *Fr* best grade of wheat flour.

farinograph — a small recording dough mixer that measures the changes in consistency of a dough during a prolonged mixing period. Its results give an indication of the quality of the flour or of its similarity to previous deliveries.

fast dough — about the same thing as an emergency dough, i.e., a dough formulated so as to ferment and condition more rapidly than a regular bread dough.

fat absorption — for a fried product, the amount of oil it takes up during the cooking process.

fat bloom — a dull white or grayish coating on a chocolate product; results from exposure to high (melting) temperatures, which allows certain fractions of the cocoabutter to come to the surface. Cannot be corrected except by remelting the chocolate, tempering it properly, then cooling.

fatier — *Egypt* appears to be a kind of non-sweet strudel made of many thin layers of dough coated with melted butter.

fat rascals — *UK* a kind of rich tea-cake made with butter or cream, and containing currants.

fats — In a chemical sense, triglycerides, i.e., the chemical species resulting from the esterification of one unit of glycerol with three units of fatty acids. A natural fat is a mixture of many different triglyceride fat molecules. Also, see "Total fat."

fatty acids — a chemical species occurring naturally, either singly or in combination with other moieties, consisting of strongly linked carbon and hydrogen atoms in a chain-like molecule. At one end of the molecule is a reactive acid group.

fer à cheval — *Fr* horseshoe-shaped loaf.

ferment — (1) To undergo the microbiologically mediated reactions that result in the production of carbon dioxide and ethanol as well as many other physical and chemical changes in yeast-leavened doughs. (2) A premix similar to a broth; a fluid mixture of flour, water, yeast, and other ingredients that can be handled by liquid transfer techniques and fermented under controlled conditions until it is suitable to be used as part of a bread dough.

fermentable solids — those materials that can be metabolized by baker's yeast to form carbon dioxide and ethanol, the most common being sucrose, glucose, fructose, and maltose.

fermentation — in baking, the complex series of microbiological changes that generate carbon dioxide and ethanol in a dough or broth. Also, the steps in the dough preparation process that are introduced to encourage or accelerate such changes.

fermentation loss — the reduction in weight occurring in a dough as the result of the change of sugars into carbon dioxide that is lost from the dough; some of the ethanol is also lost to the atmosphere, and possibly other substances escape.

fermentation room — a room or other space in which temperature, and usually humidity, can be controlled so that sponges or doughs can be uniformly fermented.

fermentation tolerance — the length of time a dough can be processed before or after the fermentation optimum and still yield good quality bread.

ferrous sulfate — an iron compound that can be used for nutritionally enriching bakery foods.

fiber — an inexactly defined group of food constituents having the unifying characteristic that they are not digested (or are poorly digested) in the human intestines, so that they contribute to the bulk of feces and encourage intestinal motility. It is not necessary that dietary fiber materials actually have the physical form of a fiber.

ficelle — *Fr* very thin loaf of bread about 10 inches long and weighing about 5 oz; a giant bread stick.

fig — fruit of the fig tree, *Ficus carica*. There are many varieties, differing in size, color, flavor, texture, etc. The American baker uses dried figs as a basis for fig bar filling, and for very little else.

figure piping — forming designs with thin strips of extruded royal icing, imitation jelly, etc. Used for decorating cakes and the like.

filberts — an approximately spherical nut generally larger than a peanut and smaller than a Brazil nut. A very common ingredient in European pastries but very uncommon in U.S.A. baked goods. Hazelnuts.

filler — (1) A mill fraction added to a blend to provide bulk or weight rather than strength. (2) Anything added to a mixture for the purpose of increasing volume or weight without significantly affecting the functionality of the mixture.

fillings — imitation whipped cream, puddings, jams, and the like that are injected into bakery products, spread between layers, or otherwise put inside semi-finished dessert items.

film — in packaging, any kind of very thin, flexible sheet made of plastic.

film gauge — the thickness of a packaging film, usually given as mils (thousandths of an inch).

filo — *Greece* (various spellings, such as fila, phyllo, fillo) a paper thin sheet of pastry, very similar in properties and usage to strudel leaves.

filter — a porous sheet used to separate solid particles from liquids or gases, or the equipment containing such a device.

final proof — the final proofing stage before the dough goes into the oven, after the piece is completely formed and in the pan (if it is a panned product). Some decorative work, such as splitting or applying wash may still be performed after the final proof and before entry into the oven.

fines lames — *Fr* assistant bakers who specialize in slitting dough.

finger rolls — buns about 5 in long and 1 in wide, made of bread dough.

finishing department — the section a bakery where the baked products come from the oven or cooler to be filled, iced, topped, coated, or otherwise made into the finished product ready to be packaged.

fire box — a combustion chamber where fuel is burned to provide heat for an oven.

fire point — the temperature at which an oil sample, when heated under a prescribed set of conditions, will take fire and burn for at least five seconds. Fire points of domestic salad oils and frying shortenings are around 650°F.

first clear — the flour portion that is collected after the patent flour mill streams have been diverted; may be further divided into fancy clear and second clear.

fiti — *Egypt* bread in pancake shape made from a batter containing wheat flour, salt, and water; cooked on an oiled griddle.

flake — one of the kinds of particles found in feed bran; also a particle of middlings material that has flattened out during grinding.

flake-buster — a device used to break up middlings flakes.

flan — (1) A tart baked in a bottomless circular metal form that has been placed on a sheet for filling and baking. (2) *Sp* a stiff custard.

flash heat — when the oven is empty but being supplied with heat, all of the materials of construction in and around the baking chamber tend to reach a higher temperature than exists when the oven is running with a full load; this heat will not normally be indicated by the thermometers, which measure only a restricted area in the baking zone, but it may cause an excessively rapid transfer of energy to the product during the first several minutes (or more) after the pans start going through the oven, i.e., before thermal input and output have stabilized. This flash heat can cause overbaking of the first loads if it is not taken into account when setting up baking conditions.

flash pans — containers of water placed in the oven to absorb flash heat.

flash point — the temperature at which an oil sample, when heated under a prescribed set of conditions will flash when a flame is passed over its surface. Flash points of typical domestic oils are around 610°F.

flat icing — a simple icing composed of water, sugar, and a stabilizer such as gelatin. Coloring and flavoring ingredients are also often added. Such icings are applied to cinnamon rolls and the like.

flavor — the combination of tastes and odors perceived when a material is ingested.

flavor extract — natural and/or synthetic materials dissolved in alcohol and intended to duplicate a natural flavor, such as raspberry.

fleurage — *Fr* dusting material (such as corn meal) applied to the bottom of dough pieces to keep them from sticking to implements such as the peel.

fleurons — *Fr* decorative shapes such as fans cut from puff pastry dough.

floor time — the period between removal of dough from the mixer and its processing by the divider.

Florentines — small, thin disc-shaped confections consisting usually of chocolate or caramel covered with nut pieces or candied fruits. Often called "cookies," but they seldom contain any flour.

flour — when not modified, the word means the finely ground material composed mostly of the endosperm of the wheat berry. There are many other kinds of flour.

flour ferment — a preferment or liquid sponge in which about 70% of the flour and all of the water, as well as other ingredients, are fermented together before they are added to the remaining ingredients.

flour streams — the many different flour-grade fractions coming off the grinding and sieving processes in a flour mill. Each stream is separately conveyed, generally in a pneumatic tube, but they can be combined in different ways to get different kinds of flour: patent, clear, straight, etc.

flour weight basis — a method of expressing the amount of ingredients in a formula by taking the amount of flour as 100%. Thus, a formula describing a batch containing 60 lbs of flour, 30 lbs of sugar, and 15 lbs of fat would be expressed as 100% flour, 50% sugar, and 25% fat. This method of formulating should generally be avoided as it seldom serves any useful purpose and it makes comparison of different formulas difficult.

flow sheet — chart indicating the machines and/or the unit operations involved in a manufacturing process, arranged in a manner illustrating their proper sequence in the line.

flue — chimney, smoke conduit.

fluff — beaten, aerated egg white and sugar to which crushed fruit has been added. Also, "fruit fluff."

fluidity — the reciprocal of viscosity.

flûte — *Fr* a loaf about 16 oz in weight and 15 to 20 in long.

flying sponge — a sponge mixture (used in a sponge-and-dough bread process) that is fermented for a relatively short period of time, usually with fermentation reactions accelerated by higher temperatures, more yeast, or other methods.

foam — any mass of finely divided bubbles that has some stability, but in baking particularly applied to an aerated mass of egg and sugar, as in sponge cake batter before the flour is added.

foam cakes — cakes based on a foam prepared by whipping egg whites and sugar; into the foam, flour and the other ingredients are folded. Angel food cake is an example.

focaccia — *It* similar to a baked pizza crust without filling, often flavored with spices mixed into the dough. Used as a bread.

focaccia di formaggio — *It* cheesecake

fold — the operation which, with doughs, consists of lapping a sheet of dough over itself, and with batters consists of lifting and turning over a portion of the mass so as to lightly incorporate ingredients (as in making angel food cake batter).

fondant — a base for icings and chocolate bonbon centers prepared by cooking sugar syrup in such a way that very small and uniform sugar crystals are formed in a continuous phase of saturated sugar syrup. The best examples appear as very white, creamy, smooth pastes.

fondant sugar — sucrose with a particle size less than about 44 microns in the largest dimension that has been mixed with a stabilizer such as maltodextrin; used as the major ingredient in certain icings, confectionery, etc.

Food Chemical Codex — a group of internationally accepted specifications for food additives.

forgácsfánk — *Hungary* a fried unleavened biscuit made of a dough rich in eggs and sour cream, but not very sweet; the dough is sheeted thin, cut into rectangles, and folded in a special way before it is fried in deep fat.

forking — also, fork splitting. An operation used to partially separate the top and bottom halves of English muffins by pushing thin blades or tines from the sides toward the center. The consumer completes the separation by manually tearing the halves apart.

form-fill-seal — a type of packaging machine that forms packets from roll stock, fills flowable food products into the packet, and seals the packet.

formula — essentially, a recipe that has been adapted for large scale use in the bakery. Contains a list of the ingredients with limited descriptions and their respective amounts or percentages. May or may not include processing instructions.

formula balance — an expression indicating that the ingredients in a formula are in the correct proportion (according to some predetermined standard) to yield a top quality product.

fougasse — *Fr* a milk roll or dough cut and formed into fancy shapes.

foxy — describes a baked product, particularly bread, that shows a tendency toward excessive redness of crust color.

fractions — substances separated from a mixture according to some specific operation; fractions themselves are often mixtures of compounds.

fractionation — in processing edible oils, means controlled crystallization and partitioning techniques using solvents, detergents, or cold dry-pressing to separate hard and soft fractions. A method for making winterized oils, high stability frying oils, and cocoabutter replacements.

frangipane — a pastry cream made with milk, sugar, flour, eggs, and butter, mixed with crushed macaroons or ground almonds; used to fill or top various cakes, pancakes, etc.

franskbrod — *Denmark* white bread.

frasage — *Fr* the initial mixing of ingredients to make dough.

free fatty acids — fatty acids that are not chemically bound to other moieties such as glycerol. The amount of free fatty acids in an oil is measured by reacting or titrating the oil with standardized alkali in alcoholic solution.

freezer burn — the loss of moisture in some parts of a frozen product due to migration of water in the vapor phase. Can occur in the absence of any melting. Usually evidenced by a change in color and texture.

French bread — an unsweetened crusty hearth bread made from a very lean dough.

French knife — a long knife with a pointed blade used for cutting cakes, doughs, and nuts.

French pastries — usually applied to an assortment of several kinds of petit fours, cream puffs, and other fancy bakery desserts, from which the customer can make a selection.

friction factor — in the calculation of the amount of cooling needed to maintain proper dough temperature during mixing, friction factor is a value representing the heat input resulting from mixing energy absorbed by the dough.

GLOSSARY OF MILLING AND BAKING TERMS

fried pie — a portion-size pie made by folding dough over a filling, sealing the edges, and frying. Usually about 4 to 6 ounces in finished weight.

fritter — a deep-fried ball of dough, often containing corn kernels, other types of vegetables, or fruits. May be served with maple syrup or the like.

front of the mill particles — relatively coarse particles removed near the beginning of the flour milling process.

frosting — an icing applied to the tops of cakes and pastries.

fructose — levulose, fruit sugar. A reducing hexose having the same composition as glucose but with a different molecular configuration. It is fermentable by bakers' yeast.

fruit cake — a term applied to a tremendous range of products, all of them characterized by their high content of candied fruits, raisins, nuts, etc. In most of them, the crumb structure that holds the fruit together in one mass is of the poundcake type, but there are many other options.

fruit pie — dessert products made by putting a mixture of fruit, sugar, and other ingredients into an unbaked pie crust and covering it either with another complete crust or with a latticework of dough, then baking it. Usually multiserving size.

fruit puree — thoroughly cooked fruit (usually of reduced moisture content) that has been strained or milled to give a paste of uniform consistency.

frumento — *It* wheat.

frying — a cooking method that relies on hot fat as the heat transfer medium. In the baking trade, usually refers to deep fat frying.

fudge — soft, creamy confections, originally made of sugar, milk, butter, and chocolate, but now having highly variable compositions due to the availability of various stabilizers, etc. A distinguishing feature is the pleasant smooth texture that results from the extreme fineness of the sugar crystals.

fudge icing — a rather dense icing usually prepared from icing sugar or fondant mixed with margarine or melted shortening.

full proof — a proof fermentation allowed to continue until the piece has reached the maximum size it can sustain without collapsing.

fumigation — applying insecticidal smoke, vapor, or gas to an room or other enclosed space.

fungal amylase — an enzyme derived commercially by growing certain types of molds in a pure culture, extracting them, and isolating the desired enzyme from the extract. Used to replace the diastatic effect of malt, etc.

fungus — a filamentous plant (often microscopic) lacking chlorophyll and reproducing by spores. Mold.

F value — in canning technology, defined as the number of minutes needed to kill a stated number of microorganisms at a specific temperature.

-G-

gajar — *Ind* carrot.
galactose — a monosaccharide that in lactose (milk sugar) is chemically combined with glucose. Not commercially available as a separate sugar.
gal-gal — *Ind* lemon.
galletas — *Sp* cookies; sometimes crackers.
galette — *Fr* a round puff-pastry loaf, about 0.5 lb, traditionally consumed on Twelfth Night.
gallon — a standard measure of volumetric capacity. The U.S. gallon is equal to four quarts of 32 fl oz each, or 3.785 liters.
galuska — *Hungary* bite-sized dumplings made of rough-textured cereals or white flour.
gansito — *Mex* chocolate-covered cake stuffed with strawberry and cream flavored filling (a trade name).
garibaldi — a cracker/cookie prepared by inclosing a thin layer of raisins or other dried fruit between two thin dough sheets, baking the combination, and cutting the sheet or strip into rectangles of uniform size.
garlic — the bulbs of the plant *Allium sativum* L., which is related to the common onion. There are white, pink, and yellow varieties, but the white type is generally favored. So far as baked products are concerned, garlic bread, breadsticks, and pizza toppings are the main applications of this spice.
gas chromatography — in this method of analysis or separation, samples are vaporized and carried by a non-reactive gas through tubes containing solid absorbent materials. Components of the sample gases are retained by the absorbent to varying degrees depending upon the extent of their chemical affinity. This results in the components exiting the distal end of the tube at different times, at which point they can be detected by various means and their quantity estimated.
gassing power — a measurement of diastatic activity in which a small amount of dough is fermented in a hermetically sealed container having a gas pressure gauge attached. The pressure rise over a period of 5 or 6 hours indicates how much fermentable sugar is being produced by amylolytic enzymes.
gastronomy — the art of good eating.
gaufrettes — *Fr* sugar wafers.
gâteau — *Fr* a specialty bread or loaf, very often of a particularly rich kind.
gehun — *Ind* wheat.
gel — a material that is firm or semi-rigid, but which can be deformed with slight force; made from appropriate liquids by cooling, adding a gum, or applying some other treatment.
gelatin — a substance obtained by extracting collagenous proteins from

animal skins or bones; used for thickening or gelling many different aqueous solutions. A distinguishing feature is the transition of its solutions from thin liquids to solid but short gels over a fairly narrow temperature range (as from room to refrigerator temperatures). Unlike pectin, it does not depend on high sugar concentrations for gelling power.

gelatinize — to cook starch in water until the granules swell and form a viscous sol.

genoese sponge — (*Fr* pâte à génoise) a very rich, air-leavened cake; contains a large amount of whole eggs and a considerable amount of melted butter.

germ — that part of the seed from which the new plant starts to develop; the embryo.

germicide — a chemical preparation that kills microorganisms.

ghee — *Ind* clarified butter, butterfat.

gill — *UK* a liquid measure, of generally, 5 fl oz, or one-fourth of an Imperial pint. Other definitions are given in the older literature.

ginger — the root (actually, in a botanical sense, the rhizome) of a plant, used either in candied form as a garnish or confection, or in the dried and ground form as a spice in many bakery products, but as a predominant flavor in only a few, such as gingerbread and ginger snaps. Has "heat" like pepper, but also has a pleasant aroma and taste.

gingerbread — includes a wide variety of products, some yeast-leavened and some chemically leavened. The former type is seldom, if ever, seen in the U.S., but they are common in Germany and some other European countries. The unifying factor in all these is the use of ginger to provide the predominant flavor. Gingercake is a less dense version of the type. American gingerbread is almost always soft, of rather high specific volume, and contains fairly large amounts of molasses.

glace fruit — fruits or pieces of fruits treated with concentrated sugar solutions so that they achieve considerable storage stability while at the same time retaining some of their original flavor and appearance.

glarus — *Swiss* flat, round bread.

glasering — or "glasyr." *Sweden* Icing or frosting.

glassine — a semi-transparent paper (not waxed) of moderate strength and stiffness.

glasur — *Denmark* sugar icing or glaze.

glaze — a transparent or translucent coating that can be applied to bakery products either as a type of icing or as a coating for the purpose of keeping exposed fruit pieces from drying out.

gliadin — a type of protein that constitutes part of the gluten fraction of wheat flour.

glucono-delta-lactone — a form of gluconic acid in which the carboxyl group is chemically combined and thus relatively inert. When it contacts water, under certain conditions the carboxyl groups are slowly exposed and become reactive. Has been used as a slow-acting leavening acid.

glucose — (1) The hexose D-glucose. (2) Corn syrup — this terminology should be avoided to prevent confusion since corn syrup contains many constituents besides the hexose in question.

glucosidic cleavage — the splitting apart of a glucose polymer, such as starch, by the chemical addition of water molecules to the glucosidic bond holding adjacent glucose units together. Facilitated by acid conditions and by enzymes.

glutathione — a reducing substance present in considerable amounts in yeast cells, and also in other natural substances. Because it can convert the disulfide bonds between protein molecules to sulfhydryl bonds (which do not join the molecules together) it can weaken gluten. This effect is useful in softening excessively strong doughs and reducing mixing times.

gluten — the mixture of proteins in wheat flour that form, upon the addition of water, the elastic structure responsible for the peculiar structure of bread doughs. There are also other proteins in wheat flour.

glutenin — one of the two kinds of proteins that form gluten, and give it strength.

gluten washing — a traditional and nearly obsolete test for flour strength; it consists of first forming a flour and water dough, then kneading the dough continuously in a stream of water so as to remove nearly all the non-gluten constituents. The remainder, containing essentially all the gluten, can be dried and weighed or otherwise evaluated.

glycerine — commercial preparations of glycerol, which see.

glycerol — chemically, a polyalcohol; a sweet, viscous, hygroscopic liquid obtained commercially as a by-product of soap manufacture; the human intestine breaks down fats into glycerol and fatty acids.

glycerol monostearate — an emulsifier and crumb softener.

glycolipid — an organic compound in which a carbohydrate moiety is chemically combined with lipid.

glycoprotein — an organic compound in which a carbohydrate moiety is combined with a protein.

Golden Syrup — *UK* a proprietary sugar syrup well known in England; described by its manufacturer as a super-saturated solution of refined sugars together with non-sugars that give it color, flavor and other properties characteristically its own.

goma — *Jap* sesame seed.

goma-abura — *Jap* sesame seed oil.

gomme — *Turkey* a flatbread made from a stiff dough containing flour and milk. The unleavened dough pieces (about 15 inches in diameter and 1 to 2 inches thick) are baked between a hot stone and a thin metal plate with hot ashes on its top.

gougère — *Fr* small puffs made from a pâte à chou mixture containing a large percentage of Gruyere cheese.

goûters fourrés — *Fr* a kind of cookie made from hard or semi-hard doughs, topped with sugar or salt and filled.

grading — (1) The practice of applying a numerical or descriptive score to grain or the like, according to an established set of specifications. (2) The tendency of grain or other material inclosed in a bin, to separate according to density and particle size. Lighter particles tend to settle near the bin wall during filling of the bin, while the heavy particles tend to exit the bin first. Can cause severe problems in dispensing if not corrected.

gradual reduction — the modern process of milling, in which the goal is to produce middlings rather than to avoid doing so, as was the principle before the New Process was introduced. In gradual reduction, the middlings are sized and separated from the bran by sieves and purifiers, and the particles in each size range are re-processed under conditions peculiarly suited to that size. The process is repeated until the desired end products are obtained.

graham crackers — present day graham crackers are usually made from chemically-leavened doughs containing a fairly high proportion of whole wheat flour and they are usually fairly sweet due to their content of molasses, etc. They are often formed in about the same shape and size as a soda cracker and can be made on the same kind of equipment.

graham flour — whole wheat flour.

grain — when used in connection with bread quality, means the size, shape, and arrangement of the cells or bubbles comprising the crumb. For most bakery products, fine and uniform cell structure with thin cell walls is the desired grain.

graining — adjusting the cooling rate and stirring conditions of a supersaturated sugar solution so that crystals of the desired size will form.

granero — *Sp* granary

granulated — formed into small, fairly uniform pieces by cutting, milling, or crystallizing under controlled conditions; not a powder; granulated sugar is an example. Nuts are often "granulated" by a cutting process.

graphite — one of the forms of elemental carbon. Can be used as a lubricant under very high temperatures where greases and oils are unstable, as on oven band supports.

gravity feed depositor — a depositor that depends on the force of gravity to draw batter into the cutting mechanism (primarily used for doughnuts).

grease — a petroleum lubricant of plastic consistency; also, a very viscous edible fat or oil.

greasing — (1) Spreading or spraying food oil or fat on dough pieces, pans, etc. (2) Applying petroleum grease to the moving parts of machinery to reduce friction and wear.

green — the property of a flour that results in a requirement for more oxidation, longer mixing, or longer fermentation than usual to avoid the production of underdeveloped doughs. Such flours tend to yield bread lacking in oven spring and having a flat top, foxy color, shiny pan crust, and coarse round cells in the crumb.

grillage — *Fr* a flavoring and texturizing material for confections made by mixing roasted nuts and caramelized sugar, then breaking down or grinding the mass after it has cooled.

grissini sottile — *It* thin dry breadsticks.

grissini conditi — *It* flavored breadsticks, a common snack item in Italy; may be flavored with fennel seeds, an oregano-pepper-basil mixture, or cheese.

grits — coarsely ground corn endosperm; also applied to other grains.

groat — what remains after the hull is removed from a grain of oats.

grzbek — *Poland* so-called "mushroom" (from appearance, not ingredients) cake, similar to kaiserschmarren.

guar — a water-absorbing and -binding vegetable gum obtained from the seed of the guar plant. Useful for thickening and stabilizing fillings, toppings, batters, etc., but not effective as a gelling agent.

guava — a tropical fruit used primarily for making jellies or purees.

gujiya — *Ind* sweet stuffed puris.

gum arabic — a powder obtained from certain types of acacia shrubs and used for increasing the viscosity of liquids or (in high concentrations or in combination with other gelling agents) for forming sticky, "long" gels.

gumming — formation and accumulation on heating surfaces of a fat-insoluble sticky material resulting from high temperature breakdown of fats and oils. The gummy material results from oxidation and polymerization of the glycerides.

gums — substances that, when mixed with water, give gelatinous, soft, rubbery masses, or, more generally, absorb water and thicken solutions. There are both natural and synthetic gums.

gum tragacanth — a natural gum suitable for increasing viscosity but not suitable for gelling purposes.

gypsum — a natural (mined) form of calcium sulfate that has been used as a mineral enrichment in bakery foods and to compensate for very soft ingredient water.

gyüszüfánk — *Hungary* deep-fried small thin discs of dough from flour, whole eggs, and salt. Called "thimble doughnuts," but mostly used for adding to soup.

-H-

hafergrütze — *Ger* oatmeal.
hafremjöl — *Sweden* oatmeal.
haldi — *Ind* turmeric
half-high grinding — see New Process
halka — *Turkey* similar to bazlama (q.v.) except baked in a peel oven.
halophilic — describes microorganisms that can grow in solutions containing a relatively high concentration of salt.
hamburger roll — a soft round bread roll of about 4 in diameter and 1.5 in thickness, made from a fairly lean yeast-leavened dough.
hard butter — a generic term used in the food industry to describe a class of specialty fats with physical characteristics similar to those of cocoa butter. They are used in confectionery coatings and centers, etc.
hard crack — a stage in sugar boiling that is reached about 280°-310°F.
har dhania — *Ind* coriander
hard red spring wheat — a type of *Triticum aestivum* from which the strongest bread bread flours used in the U.S. are milled; normally has a high content of good baking quality protein.
hard red winter wheat — the type of *Triticum aestivum* grown in greatest quantity in the U.S., and from which most of the bread flour and all-purpose flour is milled. It is not very satisfactory as a source of flour for cakes and pastries. During a single season, HRW from different sections of the country may yield flours of significantly different baking quality.
hard roll — a yeast-leavened bread roll with a crisp crust.
hard sweets — plain cookies made from a rather lean formula dough, often processed on a brake, sheeted thin, baked almost to dryness. The cookies are very crisp, or even hard, in texture.
hard water — water that contains a relatively large amount of minerals; water hardness is expresed as grains per gallon or parts per million of equivalent calcium carbonate; "Hard" water has been arbitarily defined as having 120 to 180 ppm, "Very hard" water as above 180 ppm.
harina de avena — *Sp* oatmeal.
harina de trigo — *Sp* wheat flour.
harusame — *Jap* noodles made from soybean flour.
head of mill — that part of the mill where the initial processing of the grain takes place.
hearth — the heated baking surface of the floor of the oven; the part of the oven on which the products (or pans) rest during baking.
hearth bread — originally, loaves or rolls baked on the floor of the oven, without the use of pans. Now, often applied to bread or rolls baked in or on pans that do not confine their lateral expansion.
heat exchanger — a piece of equipment that efficiently (and usually con-

tinuously) transfers heat from one fluid (liquid, gas) to another.

hedonic scale — a method for subjectively rating the hedonic quality or sensory desirability of, e.g., a foodstuff.

helote — *Mex* corn, generally sweet corn.

high amylose starch — a starch containing over 50% amylose.

high fructose corn syrup — corn syrup in which a significant portion of the glucose has been enzymatically transformed into fructose. The percentage of fructose varies between commercial products.

high milling — grinding with the rollers widely separated, necessitating a larger number of millstands. This arrangement will ordinarily yield more or better flour in succeeding grinds, but reduces the plant throughput.

high ratio cake — definitions vary, but this generally means a cake containing more sugar than flour; virtually all layer cakes made nowadays are of this type. Routine trouble-free mass production of this type of cake was made possible by the development of emulsifier shortenings and special cake flours.

high ratio cake flour — a soft wheat flour of about 8% gluten content that has had its pH lowered to 5.2 or less by chlorine treatment.

high ratio shortening — a shortening containing emulsifiers such as monoglycerides or polysorbates, and suitable for making high ratio cakes.

high speed mixer — the modern type of horizontal dough mixers with motors and gearing made strong enough to enable the mixer arms to rotate relatively rapidly; special vertical mixers are also described as such.

holding roll — the slower rotating member of a pair of rollers.

holding tank — any large vessel in which some fluid (e.g., preferment) is held for a time after it is mixed or between any processing steps.

homogenizer — a device that converts a rough mixture of fat and aqueous solution into a relatively stable suspension of tiny fat globules. Typically used to prevent the butterfat of milk from separating.

honey — the sweet viscous syrup made by bees; there are many different floral varieties, such as clover honey and orange blossom honey. The sweetening power is due mainly to fructose and glucose.

honeycake — various kinds of European specialties, generally cakes or thick cookies, supposedly sweetened only with honey. They are usually dense, rather tough, and fairly dark in color.

honig — *Ger* honey.

honung — *Sweden* honey.

hopper — a tank or receptacle, usually with angled sides, positioned above a processing device so as to receive ingredients or intermediates and hold them briefly until the processor is able to accept them.

horizontal mixer — a mixer with a U-shaped bowl having flat sides and an open (or openable) top; the agitator consists of thick metal arms running from side to side; shape of the agitator blades can vary, but is usually cylindrical for developing doughs; the inside of the bowl may contain one or more large ridges to help control dough movement. The bowl is usually jacketed so that refrigerant can be circulated around it.

horno — *Sp* oven.

hot cross buns — sweet, spicy, fruity, buns with a cross-shaped depression on top which is filled with a plain frosting. Easter specialties.

hot dog bun — a yeast-leavened bread roll with moderate amounts of enriching ingredients, sized to fit a frankfurter sausage.

hot print icer — applies hot icing to a baked product with a roller that contacts the item.

hot sponge — a method of bread baking in which the sponge is fermented at a temperature perhaps 10° to 15°F higher than used in normal practice.

hsing jen ping — *Ch* Chinese almond cookies.

huevos — *Sp* eggs.

humectant — a substance that absorbs moisture from the atmosphere; many sugars have this property. Net absorption ceases when the water activity of the substance reaches that of the surrounding atmosphere.

humidifier — a device that adds humidity (i.e., moisture) to the atmosphere of a proofing room or other space. Can be either automatically or manually controlled.

humidity — usually expressed as "Relative Humidity" which is an expression of the percent of moisture in air as related to the total moisture capacity of that air at a particular temperature.

Hungarian process — the modern flour milling system of gradual reduction with rollers and a primitive type of purifier was first put into widespread practice in Hungary, and so, for a time, such milling was called by this name.

hydrate — a compound that crystallizes with a specific number of water molecules.

hydrogenated fat — a fat that has been reacted chemically with hydrogen gas in the presence of a catalyst. The purpose is to stabilize and/or harden the original fat.

hydrogenation — the chemical operation of adding hydrogen atoms to a compound.

hydrogen ion concentration — water and all water solutions contain hydrogen ions resulting from the spontaneous ionization of water molecules; acid solutions contain more, alkaline solutions contain less, than pure water.

hydrolysis — a chemical reaction involving breakdown of molecules through their interaction with water. For example, an ester may hydrolyze in the presence of water, under certain conditions, to form an acid and an alcohol. The reaction can be catalyzed by the enzymes called lipases and by strong acids or alkalies.

hydrometer — an instrument for determing the specific gravity of a liquid; a common design is a weighted bulb surmounted by a stem or rod on which graduations have been engraved. When the instrument (which is jacketed with glass) is placed in a sugar syrup or the like, the bulb will sink in the liquid and the specific gravity (or a number that can be converted to specific gravity) is read off the stem at the surface of the liquid.

hydroponics — the practice of growing plants with their roots immersed in aqueous solutions of nutrients instead of in soil.

hydroxylated lecithin — lecithin that has been chemically treated to improve its hydration.

hydroxyl group — a chemical radical consisting of one oxygen atom combined with one hydrogen atom.

hydroxypropylmethylcellulose — a highly modified cellulose that is used to impart increased viscosity and water-retention properties to icings and the like.

hygrometer — a device for measuring relative humidity.

hygroscopic — describes a substance that absorbs and retains moisture from air that is within normal RH ranges. Glycerol and many sugars are hygroscopic, for examples.

-I-

icing — a very broad and poorly delimited category consisting of just about any sugar-containing paste that can be spread on, poured over, or otherwise distributed on the surface of a baked product. Glazes, frostings, and even some fillings overlap this category. Most adjuncts of this class are made from confectioners' sugar, fat (e.g., butter), and liquid (e.g., water), plus stabilizers, whipping agents, colors, flavors, and preservatives, as required. Meltable coatings such as chocolate and chocolate analogs are generally not considered icings but rather enrobings, couvertures, or just "coatings."

icing knife — a spatula with a handle designed for spreading icing on cakes.

icing screen — a metal screen with fairly large openings on which doughnuts can be placed while being glazed or iced; the excess glaze flows through the screen and is collected and re-used.

icing sugar — powdered sugar, confectioners' sugar (which see).

idli — *Ind* small cakes in pancake shape made from relatively coarse rice particles; cooked by steaming.

imagawa-yaki — *Jap* a chemically-leavened muffin containing bean jam.

impingement oven — oven containing many vertically-oriented tubes through which hot air is blown onto the product.

incorporating — blending one or more ingredients into a partially mixed batch.

indirect fired oven — ovens in which the basic heat source, burning fuel, is used to heat air that is then blown into the oven to bake the product.

infrared oven — such ovens transfer most of the heat energy to the product via infrared radiation; of course, all ovens (except microwave ovens) transfer some energy by way of infrared radiation.

ingrediator — a tank in which minor ingredients are blended with a small amount of water and then portions of the mix delivered to the main batches; also called a slurry tank.

injera — *Ethiopia* fermented round flatbread made from sorghum meal; the dough is fermented for 2 or more days, boiled, and steam cooked.

instore bakery — a bakery operating as a separate unit within a supermarket.

interesterification — a chemical change resulting in the random rearrangement of the fatty acids in the triglycerides of a natural fat. The distribution of fatty acid species in any naturally occurring oil conforms to a general pattern characteristic of that ingredient, but the distribution can be changed to random pattern through use of a catalyst and appropriate processing conditions. A common procedure in modern shortening manufacturing plants.

intermediate proofer — in conventional breadmaking processes, equipment that receives dough balls from the rounder and allows them to "rest" for a few minutes before they go to the molder.

intermediate proofing — a stage in which the rounded dough pieces are allowed to ferment so that the gluten relaxes, making the dough more suitable for processing in the molder.

invert sugar syrup — syrup formed by partially hydrolyzing sucrose (cane or beet sugar) into its constituent hexose sugars, glucose and fructose. Usually made by heating sucrose syrup with acid or by adding invertase to sucrose syrup.

invertase — an enzyme preparation obtained from yeast and used for hydrolyzing sucrose to give a sweeter syrup of lower viscosity and greater hygroscopicity. Also, the pure enzyme.

iodine value — an expression of the degree of unsaturation of a fat. It is measured by determining the amount of iodine that will react with a natural or processed fat under prescribed conditions. Iodine reacts with the chemically unsaturated bonds in the fatty acids.

iodized salt — sodium chloride to which a very small amount of potassium iodide has been added to provide the nutrient factor, iodine, to consumers.

iodophor — combination of iodine with another substance that causes the slow release of iodine when the combination is contacted with water.

isomers — compounds that can exist in more than one form, with different physical or chemical properties, even though they contain the same elements in the same proportions. There are two important types of isomerism: geometric and positional.

Italian bread — in the U.S., describes a loaf very similar to, indeed often indistiguishable from, French bread; hearth-baked loaves made from a lean dough containing little or no shortening, milk, or sugar.

-J-

jacket — a double-wall construction of mixing vessels, allowing circulation of a heating or cooling medium so as to adjust the temperature of the vessel's contents.

jägerbrötchen — *Ger* yeast-leavened bread rolls made from a blend of wheat flour with a relatively small proportion of rye flour.

jaggery — *Ind* crude cane sugar, unrefined.

jaibas — *Mex* half-moon shaped turnovers made from a flaky dough, filled with jam or jelly.

jaiphal — *Ind* nutmeg.

jalea — *Sp* jelly.

jam — a preparation formed by cooking large pieces of fruit (with or without seeds and usually without rinds) with a large amount of sugar, and sometimes with added pectin, until enough water is evaporated so that the residue forms a coarsely textured gel.

jäst — *Sweden* yeast.

jästpulver — *Sweden* baking powder.

jaunes d'oeuf — *Fr* egg yolks.

javitri — *Ind* mace.

jawar — *Ind* barley.

jeera — *Ind* cumin.

jelly — a semi-solid food material prepared by boiling fruit juice, sugar (or other sweeteners), and sometimes pectin and acid. Most forms of jelly are covered by federal standards of identity.

jelly crystals — a dry granular mixture containing sugar, gelatin (or some other jellifying substance), colors, flavors, etc.; used for preparing imitation jellies and jams for fillings and the like.

jelly roll — a baked product made by spreading jelly (usually imitation) on a thin layer of cake, rolling the combination into a cylinder, and cutting pieces crosswise to show the internal spiral of filling.

jelly wreath — a rolled ring of basic sweet dough containing jelly filling.

jet — *Fr* bulge in a loaf due to slitting the top before baking.

jordnötter — *Sweden* peanuts.

-K-

kabbouri — *Egypt* semicircular loaf made from a mixture of corn and wheat flours, salt, and yeast; may include enriching ingredients such as honey or egg. The dough is given a short fermentation and proof, then baked to a relatively low moisture content.

kadaifi pastry — *ME* (various spellings) a threadlike pastry made by pouring a flour and water batter through a sieve onto a griddle and then quickly sweeping off the threads before they become brown. Can be rolled around fillings of various types. Looks a little like shredded wheat.

kadin göbegi — *Turkey* a kind of hot water dough, prepared similarly to cream puff dough and fried in shallow oil to give small golden-colored round pieces with a hole in the middle. Coated with syrup after cooking.

kager — *Denmark* cakes.

kaiser roll — a bread-type hard roll resembling from the top a sort of rosette with several curved depressions radiating from the center to the edge. Now made by stamping proofed dough pieces with an appropriately shaped cutter.

kaiten-yaki — *Jap* Japanese muffin containing bean jam.

kaju — *Ind* cashew nuts.

kaka — *Sweden* cake.

kakaod — *Estonia* cocoa.

kalács — *Hungary* cakes.

kali mirch — *Ind* black pepper.

kandering — *Sweden* icing.

kanel — *Sweden* cinnamon.

káposztás pogácsa — *Hungary* biscuits containing fried cabbage.

karask pärmiga — *Estonia* barley bread.

karaya gum — a natural gum that swells to form a viscous gel when heated in water. Used to thicken and stabilize toppings and other bakery adjuncts.

kardemumma — *Sweden* cardamom.

karintou — *Jap* cookie prepared by frying (a) a yeast-fermented dough made from hard wheat flour or (b) a chemically-leavened dough made from soft wheat flour.

kása — *Hungary* porridge-like food made principally of grains such as millet, with flavoring and enriching ingredients such as meat.

käsekuchen — *Ger* cheesecake.

kawara-senbei — *Jap* a tile-shaped wheat flour cracknel.

keedetud koogid — *Estonia* a kind of fried cookie made from a chemically leavened dough, usually flavored with lemon or vanilla.

keik me karitha — *Greece* coconut cake.

keks — *Ger* cookies and crackers.

kenyér — *Hungary* bread.
kernel — the seed of cereals, nuts, etc.
kernel paste — apricot kernels ground to a very fine particle size with sugar. Used in the same way as almond paste, as a texturizing and flavoring material. Very common ingredient of fillings for sweet goods such as coffee cakes. Although regarded by many as merely a cheaper substitute for almond paste, it does have an attractive flavor of its own.
khameri roti — *India* a sponge made from whole wheat flour, yogurt, salt and sugar is fermented overnight, then doughed-up with additional flour, baking soda, and water.
khejoor — *Ind* dates.
khubz — *ME* a yeast-leavened loaf resembling pita bread.
kirsch — a liquor distilled from fermented cherry juice. Occasionally used as a flavoring for gourmet bakery foods, but more often found as an ingredient in continental confections.
kishmish — *Ind* raisins.
kisra — *Sudan* flatbread made from a thick paste of sorghum flour, water, and sourdough that has been fermented about 14 hours. Baked for less than a minute on an oiled metal or clay surface. A minor amount of wheat or millet flour is often included in the dough.
kiss — a deposit of meringue in a kind of mound, baked slowly until it is almost dry. Generally, browning is to be avoided.
kis sütemény — *Hungary* teacakes or cookies.
kjeldahl — a procedure for determining total combined nitrogen in foods. The nitrogen so determined can be multiplied by a factor, such as 5.7 for flour, which gives the approximate protein content of the material.
knafi — *ME* see "kadaifi."
knead — to work a dough without cutting or tearing it, with the intent of developing the dough and/or releasing excess gas.
knock-back — releasing excess gas from a fermenting mass of dough by any kind of manipulation or mechanical treatment that accomplishes this purpose.
kolaches — many different formulas and procedures have been published for this Czechoslovakian pastry/cookie. In its original form, it was a small yeast-leavened sweet bun with a minor amount of jam or some other type of fruit filling centered on the top. Among the many variations are chemically-leavened pastry and cookie types.
kohupiima korp — *Estonia* cheese cake.
kollane pühadesai — *Estonia* Estonian Easter cake; yeast leavened, with saffron and orange peel, braided dough.
köömneid — *Estonia* caraway seeds.
korngryn — *Sweden* barley.
korsan — *Arabia* disk-shaped flatbread of about 2 oz piece weight made

from whole wheat flour, water, and salt. Given two short fermentation periods and one proof period. Thin sheets baked almost to dryness.

kosher — a food product conforming to Jewish dietary laws.

kourambiethes — *Greece* almond shortbread biscuits.

kransekagemasse — *Denmark* marzipan.

kringler — *Denmark* pretzels, including sweet doughs in pretzel shape.

krokant — almond pieces mixed with molten caramelized sugar and, when cooled, broken into small pieces or ground. Used to flavor confections, etc.

kryddpeppar — *Sweden* allspice.

kuchen — *Ger* coffee cakes and the like made of yeast doughs.

kugelhopf — an Austrian sweet yeast cake containing raisins or currants and baked in a special high crownlike pan.

kümmel — a liqueur flavored with cumin and caraway. Infrequently used to flavor bakery goods; more often used in European confectionery.

kuri-manju — *Jap* chestnut bun with bean jam filling, chemically-leavened.

-L-

label — a piece of paper or plastic in or on the container of a product.

labeling — the printed information on the package or label or otherwise accompanying a product.

lactalbumin — a protein that remains dissolved in the whey when casein is precipitated from milk. It can be coagulated by heat.

lactase — an enzyme that splits lactose into the monosaccharides galactose and glucose.

lactic acid — the principal acid in sour dairy products; it is formed by certain bacteria acting on lactose and other sugars; it is a component of the flavor formed by rye sours.

lactic acid bacteria — microorganisms that can ferment lactose, forming lactic acid and other materials in the process. They are important contributors to flavor in most types of sourdough breads.

lactoglobulin — a water-soluble protein found in milk and whey.

lactose — milk sugar; it is not fermentable by yeast and has low sweetness. Chemically, it is described as a disaccharide containing glucose and galactose residues.

Lady Baltimore cake — rich white layer cake with fruit and nut filling and white icing.

lady fingers — oblong sponge cakes typically about 1 inch wide, 3.5 inches long, and 0.3 inch thick. Most examples have a very light texture, are high in egg content, and are flavored with vanilla.

lahmacun — *Turkey* a yeast-leavened dough that has been made into a disc by pressing and rolling, then covered with a tomato paste concoction before being baked. Similar in many ways to pizza.

lame — *Fr* a special knife or blade used to cut slits in dough pieces.

laminating — forming layers of dough or alternating layers of dough and fat. An important step in making soda crackers, puff pastry, etc.

languette — *Fr* the central part of a croissant.

lard — rendered hog fat.

latent heat — heat taken up when a substance changes its state, as when water changes to steam or to ice without showing a change in temperature. Contrasted to sensible heat, which is heat that causes a change in temperature of the affected material.

layer cake — although this designation would generally apply to any kind of cake baked in layers, the term is largely restricted in practice to the high ratio cakes without nut or fruit ingredients that are iced or frosted whether in single or multiple layers.

lahvosh — *ME* (various spellings) round and crisp cracker or bread made of wheat flour, malt, sesame seeds, and yeast (or sourdough); very thin and from small to large diameter.

lauric fats — these ingredients typically contain 40 to 50% lauric acid in combination with lesser amounts of other relatively low molecular weight fatty acids. Lauric fats are obtained from the fruits of various types of oil palms, such as coconuts and palm kernels.

lean doughs — doughs made primarily with flour, water, salt, yeast, and perhaps malt but with little or no enriching ingredients such as shortening, milk, and sugar.

leaven — originally, a portion of dough saved from day to day and used to inoculate a new batch of dough so it would ferment properly; now, the word is occasionally used as a short form of "leavening."

leavening — anything used to generate gas inside a dough so as to provide the typical internal structure of a baked product. Yeast, baking powder, and ammonium bicarbonate are the most common leaveners. Water vapor and expanding air also contribute to the leavening of bakery products.

leavening agents — those substances or materials (including steam and air as well as yeast and baking powder) that cause an increase in volume of a dough or batter.

lebkuchen — a dense, firm-textured European cake (or cookie) made with rye flour and honey, and highly spiced. Usually contains glazed fruits. Many variations exist, some of them characteristic of certain localities, such as Bremen pepper cake, Baseler lebkuchen, and Nürnburger lebkuchen.

leche — *Sp* milk

lecithin — chemically, applies to certain kinds of phosphatides occurring in both plants and animals, but the food ingredient "lecithin" is almost always obtained from crude soybean oil. Widely used in foods as an emulsifier and wetting agent.

lefse — *Sweden* flat bread made from potatoes or potato flour, wheat flour, water, salt, etc.

legumbres — *Sp* vegetables.

lehetainas — *Estonia* puff pastry.

lemon curd *UK* a kind of lemon custard, often made without milk. Sometimes used as a tart filling.

levadura en polvo *Sp* baking powder.

levulose — fructose.

levure — *Fr* yeast.

levure en poudre — *Fr* baking powder.

light meal — the ground-up scrap of light-colored and mildly flavored cookies that have been rejected or returned for some reason. Used at low levels as a bulking or non-characterizing ingredient in cookie doughs.

lima — *Sp* lime (the fruit).

lime — (1) A citrus fruit, typically green and somewhat sourer and smaller than the average lemon; Mexican, Persian, and Key are three commercial types of limes. Key lime pies are considered great delicacies although most of them get their flavor from plants in New Jersey. (2) The mineral substances calcium oxide (unslaked lime), calcium hydroxide (slaked lime), or calcium carbonate (air-slaked lime), some of which have possibilities as dietary calcium supplements.

limón — *Sp* lemon.

linkage — the specific arrangement by which atoms, or combinations of atoms, are joined together to form molecules.

Lintner value — a figure obtained by determining the rate at which malt can produce reducing sugars from soluble starch under defined conditions. For bread bakers, it is an important specification for malt preparations.

Linzer torte — a rather thin, usually circular, layer of shortbread made with high levels of butter and filled with raspberry jam.

lipase — an enzyme that breaks fats into free fatty acids and glycerol.

lipoxidase — an enzyme that acts on lipids and certain related compounds. As a baking ingredient, lipoxidase is used to oxidize the yellow pigment of flour, thereby lightening the color of the dough and the finished bread.

liqueur — a compound of ethanol (often in the form of raw brandy), sweeteners, flavors, colors, and other ingredients. Liqueurs can be used as flavors in bakery products, but are very expensive for this purpose.

liquid sponge — a sponge or pre-ferment containing flour, made with enough water so that the finished intermediate can be transferred, stored, and measured by liquid handling procedures.

liquid sugar — a syrup particularly suited for bulk handling techniques, usually consisting of a 67% aqueous solution of beet or cane sugars; these syrups almost always contain a small amount of invert sugar.

loaf cake — cake batter that has been baked in bread pans or similar deep containers.

loaf volume — the volume (usually expressed in cc) of a loaf of bread.

locust bean gum — carob gum, obtained from the seed of the locust bean tree. Used to increase viscosity and retain moisture in icings, fillings, and batters. Does not normally form gels.

long patent — a patent flour that includes a relatively large percentage of the total flour produced by a mill, sometimes up to 95% of the total flour.

long system — a milling system involving a relatively large number of reduction processes; it leads to perhaps 15 different products.

lönnsirap — *Sweden* maple syrup.

loukoumathes — *Greece* honey puffs; fried yeast-leavened dough in small pieces, dipped in hot honey immediately after frying.

louver — a slanted slat arranged (usually in multiples) to allow air to flow in or out of a chamber, as in a dryer; often adjustable to control the rate of air flow. "Louvre" is an incorrect spelling of this word.

Lovibond color — color of, e.g., oil measured by a technique utilizing a series of red and yellow glass slides as standards for visually matching the color of the oil sample. Colors so measured are usually reported in numerical values of red and yellow. The Lovibond scale and associated equipment is an official measuring device of the American Oil Chemists Society.

low milling — grinding grain with rollers set close together, a method in general use before the so-called New Process became popular. The aim was to produce as much flour as possible with one grinding step.

lukier — *Poland* icing.

-M-

mabatt — *Egypt* same as kabbouri, but the loaf is larger in diameter.

macaroon paste — a combination of almond paste and kernel paste.

macaroons — these confections were originally made of ground or chopped almonds mixed with two parts of sugar and bound with egg white, but now macaroons are more likely to be a kind of coconut cookie including some flour.

mace — a spice that comes from the same tree as the nutmeg. It is the orange colored fibrous material surrounding the kernel of a fruit, the kernel being nutmeg. Generally has a milder and more "rounded" flavor than nutmeg.

MacMichael viscometer — a rotating viscosimeter used to measure the viscosity of soft wheat flour mixed with dilute lactic acid. The results are often useful in evaluating the suitability of such flour for a particular application.

mahlakas rulltort — *Estonia* jelly roll.

maicena — *Mex* corn flour or corn meal.

Maillard reaction — the so-called "non-enzymic browning reaction" resulting when amino acids and reducing sugars react at high temperatures; it produces poorly defined brown-colored compounds that sometimes have objectionable aroma and taste,

maiz — *Mex* corn (the dry field corn).

maize — corn, Indian corn, the grain from *Zea mays*.

majsmjöl — *Sweden* corn meal.

makagigi — *Poland* a confection of almonds cooked in honey and sugar, then cut into strips. Something like nut brittle.

make-up — the manual or mechanical manipulation of dough required to prepare the shape that will enter the oven.

make-up time — the duration of the period required to process the dough from the end of bulk fermentation to panning.

makkai — *Ind* corn.

makkai ki atta — *Ind* corn flour/meal.

mákosgubó — *Hungary* a curious dessert prepared from small balls of rich yeast dough by first baking until slightly brown, then briefly stirring in boiling water, and finally heating in hot oil. Usually covered with poppy seeds and sugar.

makron — *Sweden* macaroons.

maloug — *Yemen* thin, disk-shaped flatbread made from wheat flour, water, salt, and yeast. Dough fermented about two hours, then shaped by pulling, and baked inside a clay oven.

malt — barley that has been allowed to sprout under controlled conditions, and then is de-sprouted, dried or roasted, and ground. Wheat, and perhaps other grains, can be processed in somewhat the same way. Valuable as a source of flavor, color, fermentables, and enzymes.

malta — *Sp* malt.

maltase — an enzyme found in yeast, and elsewhere, which converts maltose into glucose.

maltose — a disaccharide formed from the chemical union of two glucose units; not quite as sweet as sucrose, but more hygroscopic. It is a reducing sugar.

maltose value — a measurement of the amylase activity of flour; the amount of reducing sugar produced in one hour from 10 gm flour held under standard conditions is reported as mg of maltose.

malt syrup — an ingredient prepared by extracting barley malt with water and evaporating the resulting solution. High in maltose and may have significant enzyme activity.

mandel — *Sweden* almonds.

mandeln — *Ger* almonds.

mandoletti — *Ital* a confection consisting of white nougat, fruit, and egg whites.

manju — *Jap* generic term for Japanese buns with or without filling.

mano — *Mex* a stone cylinder, generally less than a foot long, often made with tapered ends so that it resembles a spindle. Used with a metate to grind nixtamal.

manteca *Sp* lard or other cooking fat.

mantecadas — *Sp* small cakes of the kind sold in frilled papers.

mantequilla — *Sp* butter.

mantequilla de cacahuates — *Sp* peanut butter.

manzana — *Sp* apple.

maple syrup — sap of the sugar maple tree that has been concentrated by heat evaporation.

maraschino cherries — artificially colored and flavored cherries. Commercial goods of this name are no longer flavored with liqueur, at least not in the U.S.

marble cake — cake in which two or three differently colored layer cake batters have been combined in swirls before baking.

margarine — a plastic or flowable emulsion serving as a substitute for butter and containing a minimum of 80% fat. The non-fatty portion consists of some combination of water, milk products, salt, color, emulsifiers, flavor, and other additives.

marrons — sweet chestnuts, available as a puree, candied, or dried. Common as a flavoring and texturizing ingredient in Europe, but rare in U.S.A.

marrons glacè — candied sweet chestnuts.

marshmallow — a soft confection foam made of whipped sugar syrup and corn syrup stabilized with gelatin and other gums.

marzipan — almond paste mixed with powdered sugar and corn syrup. The term is also applied (at least in the U.S.) to the fruit, vegetable, and flower shapes formed from colored marzipan and used either as separate confections or added to baked products as decorations.

masa — the raw material for tortillas and similar products; made by cooking corn kernels in a slightly alkaline solution, washing (sometimes), and grinding to a paste. More generally, any kind of dough.

masa trigo — *Mex* a dry mixture for making the wheat flour type of tortilla.

mash — in brewing, the programmed heating of a mixture of malt, water, and other materials for the purpose of extracting the maximum amount of desirable fermentables.

masking — covering with icing or frosting.

masoor dal — *Ind* lentils.

maturing — allowing flour to age to improve its processing qualities, or adding chemicals to accomplish the same results.

matzo — a thin, unleavened bread acceptable for consumption during the Jewish Passover. Usually made from a low absorption flour and water dough which is developed by repeated sheeting and laminating before being docked and baked in a hot oven for about a minute. Has acquired some popularity as a snack food, leading to a proliferation of varieties.

mazurek — *Poland* cakes or tortes made with whipped egg whites and a high proportion of ground almonds. Many flavors and forms.

meal — coarsely ground grain.

medium chain triglycerides (MCT) — triglycerides (fats) containing C6, C8, and C10 saturated fatty acids. They are absorbed quickly by the body and are transported via the portal system. Conventional fats and oils containing C16 and C18 fatty acids are absorbed more slowly and are transported by the lymphatic system.

mehl — *Ger* wheat flour.

mejorana — *Sp* margarine.

melass — *Sweden* molasses.

melassa — *It* molasses.

mélasse — *Fr* molasses.

melasse — *Ger* molasses.

melaza — *Sp* also, "melote." molasses.

melba toast — thin slices of bread toasted slowly to dryness and until they assume a uniform light brown color.

melocotón — *Mex* peach.

melting point — the temperature at which a solid becomes a liquid.

menrui — *Jap* noodles (generic).

meringue — egg whites beaten with sugar; usually a small amount of acid (such as cream of tartar or lemon juice) is added to improve volume, stability, and color. If a flavor is added, it is usually vanilla.

meringues — shapes formed from meringue and then baked almost to dryness; served either as separate confections or used as a base for other mixtures.

mermelada — *Sp* fruit jam or preserves.

mesh — the number of wires (or threads) per inch in a sieve; can be expressed in different systems of measurements.

metabolism — the sum of the chemical reactions that make up the life processes in organisms.

metallized films — plastic films that have been covered with a very thin layer of some metal (usually aluminum) by vapor deposition or other methods.

metate — *Mex* a three-legged stone utensil with a flat, sloping surface; used in conjunction with a mano for grinding nixtamal in making tortillas. Generally, the grinding surface is somewhat concave.

methylcellulose — a cellulose gum in which many of the hydroxyl groups have been replaced by methoxyl groups. This treatment normally leads to a "stronger" gum.

metric system — a decimal (i.e., scaled by tens) system of weights and measures, with the meter and gram as bases of length and mass, respectively.

mett — *Estonia* honey.

microorganisms — in general, independently living organisms that are too small to be seen with the naked eye; more specifically, yeasts, bacteria, molds, and the like.

microwave oven — an oven that relies on the transformation of radiant energy of specific wavelengths to heat within the product to be cooked.

middlings — the larger particles coming from the floury part (endosperm) of the wheat berry during milling; some small bits of bran may be present. In the animal feed trade, "middlings" means fairly high grades of animal feed, i.e., mixtures of bran, germ, some of the large endosperm chunks, etc.

middlings rolls — roll stands the operation of which produces mainly middlings; sometimes called reduction rolls.

mie — *Fr* the crumb of bread, i.e., everything but the crust.

miel — *Fr* and *Sp* honey.

miele — *It* honey.

miga — *Sp* the crumb of bread, i.e., the part not crust.

mil — one thousandth of an inch, a measurement much used in the packaging film industry.

mildew — growths of fungi (perhaps other organisms as well) on food products and other materials, leading to the production of white areas and to other defects.

milk bread — white bread in which there is an amount of milk solids specified by federal standards.

milk sugar — lactose, q.v.

milles feuilles — *Fr* puff pastry in general or napoleons in particular.

millet — a name that has been applied to a wide variety of cereal grains, even sorghum, but is more properly used for two tribes of the grass family, the Chlorideae and the Paniceae. The former includes African ragi (finger millet) as the only species of economic importance, while the latter tribe includes several species grown for food and feed in various parts of the world. *Panicum millaceum*, proso or common millet, is probably the only genus grown to any extent in the U.S.

millfeed — a milling by-product, such as bran; any product of a mill not identifiable as flour.

milpa — *Sp* cornfield.

mince meat — a cooked mixture of raisins, chopped apples, candied fruits, spices, beef suet, etc. Used as a filling for pies. At one time, contained a fairly large percentage of finely chopped beef, but not any longer.

mineral oils — hydrocarbons derived from petroleum; used for lubrication purposes in the bakery and, sometimes, as part of the pan grease mixture.

mint — flavor derived from the leaves of one or more of the Menthus plants; spearmint and peppermint. Synthetic oils chemically identical to the natural essences are also available. Sometimes used in the fillings for sandwich cookies.

mitt — an insulated glove or mitten used for handling hot objects.

mixing bowl — the container affixed (permanently or temporarily) to a mechanical mixer for holding the ingredients as they are being stirred; also, a large hemispherical bowl used for manual blending of ingredients.

mixing tolerance — the relative capacity of a dough to withstand changes in mixing conditions, especially variations in mixing times.

mixing tolerance index — a figure determined by measurements on the curve drawn by a Brabender farinograph. Larger figures indicate the likelihood of greater mixing tolerance during production.

mjölk — *Sweden* milk.

mlinci — a thin flatbread made in Croatia; similar to chapatis in preparation method.

mocha — originally, a special kind of coffee bean. Now, usually means a blend of chocolate and coffee flavors.

moisture and volatile matter content — determined as the weight lost by a food material after it has been heated for a prescribed time under controlled conditions. Reported as percent of the original weight of the sample.

moisture vapor transmission rate — (MVTR) the rate at which water vapor passes through a packaging film, or other material, as determined under controlled conditions.

molasses — as a bakery ingredient, refers either to a syrup obtained as a by-product in the refining of cane sugar, or made as the prime product by evaporating sugar cane syrup. Colors of these products range from light tan to almost black, with flavor (acidity, bitterness, metallic) becoming stronger as the color becomes darker. Used as a coloring and flavoring ingredient in many bakery products.

molcajete — *Mex* a mortar shaped grinding utensil consisting of a hemispherical hollow in a stone base, used with a tejolote to grind the nixtamal used for tortillas.

mold — (1) Hollow forms of plastic, metal, or (rarely) ceramics used to shape confections; if used for baking, they are commonly referred to as "pans." (2) A fungus, i.e., a mycelial microorganism.

molder — also, spelled "moulder." A machine that shapes dough pieces for baking.

molletes — *Sp* a circular sweet bun with colored paste forming a circle with four rays on top.

mold inhibitor — a chemical substance that can be added to a formula to delay fungal spoilage; also, can be sprayed on the surface of, e.g., English muffins.

mon-le-bway — *Burma* fried rice flour batter made into crisp, thin sheets that are very light due to the presence of innumerable air bubbles.

monocalcium phosphate — also called calcium acid phosphate or acid calcium phosphate. A mineral substance used as a leavening acid, yeast nutrient, and dough acidulant.

monoglyceride — a chemical compound formed by the combination of one fatty acid unit with one glycerol unit. Used as surfactants and to delay texture staling of bread.

monorail proofer — a conveyor consisting of a single rail serving as the track for wheels from which are suspended racks containing the dough products.

monosaccharide — a simple sugar containing 3 to 9 carbon atoms (6 carbons in all ingredient sugars) with the same number of oxygen atoms and twice that number of hydrogen atoms; more complex sugars are made up of combinations of monosaccharides.

monosodium glutamate — an essentially flavorless compound that acts as a flavor enhancer or appetite stimulator.

moongfalli — *Ind* peanuts.

mousse — originally, a frozen dessert of sweetened and flavored whipped cream with fruit pieces, often with gelatine as a stiffener; the term is now frequently used for a whipped dessert high in fat, flavored (most often with chocolate) but not always frozen. Sometimes, even used for whipped starch puddings.

mqdrzki — *Poland* fried cookies or pastries made of pot cheese, eggs, sugar, and flour. Aerated only by beaten eggs.

muffins — chemically-leavened batters baked in the small cups of muffin

pans, usually fairly sweet. Also, yeast-leavened soft doughs cooked in rings on a hearth, e.g., English muffins.
multigrain breads — breads that contain substantial amounts of other grains in addition to wheat (or wheat flour).
muna — *Estonia* eggs.
munakook — *Estonia* sponge cake.
munavalgekoogid — *Estonia* meringue cookies made with egg whites, sugar, and finely chopped almonds.
mushi-manju — *Jap* a steamed bun, chemically leavened and containing a bean jam filling.
mushimono — *Jap* steamed food (generic term).
muskott — *Sweden* nutmeg.
mycotoxin — a poison elaborated by fungi.
mylar — a polyester resin used to make films for food packages.

-N-

nalesniki — *Poland* crepes or pancakes.

nan — *Ind* (also, "naan") a flatbread made from flour, yeast, sugar, salt, water, yogurt, and shortening. Although usually yeast-leavened, chemically leavened versions are known. Similar in size and shape to chapatis.

napoleons — sheets of baked puff pastry alternating with cream filling, cut into rectangles and sometimes iced.

naranja — *Sp* orange (the fruit).

nariyal — *Ind* coconut.

naspati — *Ind* pear.

nata — *Sp* cream.

natillas — *Sp* custard.

nesselrode pie — rum-flavored Bavarian cream pie filling mixed with assorted preserved fruits and placed in a pre-baked pie crust; traditionally topped with chocolate curls.

neufchâtel cheese — a soft cheese somewhat similar to cream cheese but lower in fat and therefore cheaper and lower in calories. Has been used as a replacement for cream cheese in some pastry fillings.

neutral detergent fiber — one form of food fiber, perhaps more accurately defined as one method of determining certain types of food fiber; gives higher results than the traditional crude fiber determination.

neutralizing value — a number describing the strength or power of a leavening acid, being the number of pounds of sodium bicarbonate required to neutralize 100 pounds of the acid.

New Process — the process of high grinding with millstones, using one or more lower regrinds, that came into use in the U.S. just before the introduction of modern roller milling. In comparison with very long continental systems, those procedures favored in the U. S. were sometimes described as "half-high grinding."

niacin — a water soluble (B) vitamin; one of the nutrients required to be added to enriched flour.

nimbu — *Ind* lime.

nip — as applied to a pair of mill rolls, the minimum distance between the corrugations of the cylinders as they rotate.

nixtamal — the soaked, cooked, unground corn kernels used to make masa. An intermediate in the preparation of tortillas.

nondiastatic malt — malt syrup (wet or dried) that has been heat treated to such an extent that very little amylolytic activity remains.

no-time dough — a straight dough, which through the use of more fermenting agents and higher temperatures than normal, and usually with the aid of more mechanical development in the form of mixing, has its fermentation period reduced from hours to less than about 20 min. These doughs are sent to make-up immediately after mixing, with a generally unregulated floor time during which some fermentation occurs.

nötter — *Sweden* nuts.

nougat — originally, a candy made by stirring nuts (such as almonds or pistachios) into molten sugar, then adding a fatty ingredient to soften the texture. Now, usually, a slightly aerated mixture of sugar syrup, egg white, and condensed milk. Other mixtures are also being called nougat.

nueces de acachù — *Mex* cashew nuts.

nueces de Brasil — *Mex* Brazil nuts.

nueces de Castilla — *Mex* walnuts.

nuez — *Sp* walnut; *Mex* pecan.

Nulomoline — a trade name for a standardized invert sugar syrup.

nutmeg — a spice frequently used in bakery products, made from the ground or grated kernel of an East Indian tree.

nutrient — substance in food or drink that can be taken up by the body and used as a metabolite.

-O-

oatmeal — various chopped or ground forms of the grain of the oat plant, sometimes lightly toasted.
oats — the seed of the cereal plant *Avena sativa*.
odrajahu — *Estonia* barley flour.
oeufs — *Fr* eggs.
offal — an English term for the millfeed fractions coming from a flour mill.
oil — in food processing, a natural or processed edible fat (triglyceride) that is liquid under normal storage or usage conditions.
oiled dough — dough pieces that have had oil applied to their surfaces, generally immediately before baking; the usual purpose is to eliminate the necessity for greasing the pans, although other benefits are sometimes obtained,
oil-fired oven — an oven that derives its heat from the combustion of atomized fuel oil.
oil seeds — any plant seeds from which it is commercially possible to extract edible oil; soybeans and cottonseeds are examples.
o-kashi — *Jap* cakes in the broadest sense, not necessarily leavened or even baked. The predominant ingredient will be mashed (sweet) potato or mashed beans cooked with sugar, isinglass, seaweed, and sometimes sake; molded into blocks or fanciful designs.
oklablomos — *Greece* a round flatbread.
old dough — yeast dough that has become sour and weak from overfermentation, having been held too long or at too high a temperature for the formula in use. These doughs yield loaves characterized by boldness, rough pan crust, dull top crust, irregular collapsed cell structures, and circular whorls in the crumb. The loaves may also be dark, sour, and of low volume.
oleo — also, oleo oil. Liquid edible beef fat, rendered from tallow and the like.
olive oil — oil pressed from the fruit of the olive tree. It is high in unsaturated fat and has a characteristic flavor that is much liked by some consumers and disliked by others.
omochi — *Jap* rice cake.
opacity — degree of opaqueness, resistance to transmittance of light.
organoleptic — external stimuli detectable by the senses, e.g., flavor, odor, color, etc.
orus — *Egypt* a yeast-leavened breakfast pastry consisting of a rich but not sweet dough folded over a chopped date filling which has been fermented overnight, then baked.
osmophilic — describes microorganisms that can function in systems having a low water activity (jellies, high solids syrups, etc.).
ostkaka — *Sweden* cheese cake.

othello — confection made from a highly aerated mixture of egg whites, egg yolks, and sweetener that is formed into small mounds and baked on paper. Sometimes coated with chocolate after cooling.

ounakook — *Estonia* apple cake.

ounamungad — *Estonia* similar to apple dumplings, but the apples are wrapped in a lean yeast-leavened dough.

oven — a heated chamber or partially inclosed space used for cooking foodstuffs other than by a liquid heat transfer medium. Many forms and sizes are in use.

oven hook — a hook with a long stem attached to a wooden handle; used for reaching into a hot oven to remove pans.

oven sheet — a recording form on which the oven operator enters important data regarding conditions affecting the baking product.

oven spring — the expansion of the loaf that occurs during baking.

over-and-under scale — a weighing device on which the indicator shows the weight as being either over or under a pre-set figure. Useful for rapid on-line determination of weight compliance.

overhead proofer — equipment for holding and transporting dough pieces (usually through an inclosed chamber) during the intermediate proofing period; so-called because it is usually installed near the ceiling to make the best use of bakery space.

overproof — deterioration of dough resulting from excessive proof time and/or temperature.

oxidation — a chemical reaction involving the addition or combination of oxygen with another material, or, more generally, an increase in the number of positive charges on an atom or a reduction in the number of negative charges. Oxidation in food products containing fat can result in the development of rancidity with accompanying objectionable flavors and odors.

oxidative rancidity — the unpleasant smell, often accompanied by a bad taste, that occurs when a fat has been oxidized, i.e., when the fatty acids have acquired oxygen atoms at their double bonds.

oxygen scavenger — a material that can absorb or otherwise remove traces of oxygen from a hermetically-sealed container.

-P-

paddle beater — a mixer agitator of the batter beater type.
pähklitort vahukoorega — *Estonia* walnut torte
pain — *Fr* bread; a loaf.
pain complet — *Fr* whole wheat bread.
pain de campagne — *Fr* country-style bread, usually a round hearth baked loaf weighing several pounds.
pain de gruau — *Fr* white bread made with the very best available flour.
pain de ménage — *Fr* family loaf, or regular bread.
pain de mie — *Fr* loaf baked in an inclosed pan (a la Pullman) to produce a loaf of closely defined dimensions and cross-section.
pain ordinaire — *Fr* common or standard bread.
palacsinták — *Hungary* pancakes, usually served with a filling.
palacsinta tészta — *Hungary* pancake batter.
palette knife — a spatula, generally thin and flexible with a rounded end, used for spreading icings and the like on cakes and other baked products.
pallet — a platform generally made of a cheap grade of lumber, having two layers of boards between which the forks of a lift truck can fit. They are used as bases on which containers of ingredients and products can be stacked for storage or transfer. Some are made of plastic.
palm oil — oil pressed from the fleshy part of the fruit of the oil palm.
palm nut oil — oil obtained from the kernel or seed of the fruit of the oil palm.
pan — *Sp* bread.
panaderia — *Sp* bakery.
panadero — *Sp* baker.
pan bread — loaves baked in pans or tins, as contrasted to hearth bread.
pan centeno — *Sp* brown bread; rye bread.
pan crust — a part of the crust that has come in direct contact with the inside surface of the pan during baking.
pan de caja — *Mex* loaf of bread which has been baked in a rectangular pan.
pandowdy — a dessert dish consisting of sweetened and spiced apple slices that have been covered with deposits of rich biscuit dough or streusel, then baked.
pan dulce — *Mex* sweet dough buns in various flavors and shapes; usually they do not have icings or other decoration but the dough may be colored.
panetone — *Mex* a simpler, less rich, version of the Italian panettone; a pastry containing candied fruits.
panettone — a rich yeast bread of traditional shape containing fruits; an Italian Easter specialty. The original method of processing is very elaborate and time-consuming.

pan fino — *Sp* generic term for Mexican-style pastries made of a fairly lean dough and decorated with a baked-on paste made of flour, sugar, water, shortening, colors, and flavors.

pan francés — *Mex* hearth-baked loaf bread, made from lean dough, usually having a split top.

pan grease — generally, compounds of food oils and/or mineral oils with various additives; used for coating the insides of pans so the baked products can be easily removed.

pan liners — shaped forms of paper or parchment, sometimes treated with non-stick additives such as silicones, to be inserted in baking pans for the purposes of facilitating product removal and assisting in maintaining product integrity during transfer.

panning — the process of placing dough pieces in baking pans, manually or by machines.

pannkoogid — *Estonia* pancakes.

panocha — *Mex* a circular loaf of bread, perhaps 10 or 12 inches in diameter and about 1 or 2 inches thick, usually made from a lean chemically leavened formula and cooked in a skillet over an open fire; traditionally made on the trail by cowboys, deer hunters, and other campers.

panque nuez — *Mex* cake-shaped sweet bread topped with pecans.

panqueque — *Mex* pancake.

pan rack — a stationary or movable structure of open shelves on which baking sheets or pans may be placed.

pans — variously shaped metal containers for baking or cooking.

pan tostado — *Sp* sweet toasted white bread.

pan washer — a specialized automatic washing device for cleaning pans and similar utensils.

pao — *Chi* steamed Chinese bun.

papa — *Mex* potato.

papain — a protein-digesting enzyme obtained from papaya fruit; it is much used in meat tenderizers and has been used as a gluten softener, but has been displaced from the latter application by fungal enzymes.

pappadams — *Ind* round flatbread, thin but of fairly large diameter, made mostly from lentils, sometimes flavored with cumin and/or pepper. Fried in shallow oil.

paratha — *Ind* an unleavened dough ball made with whole wheat meal that has been coated with fat (ghee), then flattened and fried on a hot iron plate.

parfait — ice cream and syrup (or whipped cream) placed in alternate layers in a serving glass. Also used fancifully to describe striped, filled hard candies.

Parker House rolls — bread rolls of about 2 oz weight formed by folding 1/3 to 1/2 of a rectangular dough strip over the rest of strip.

parmesan — a hard Italian cheese with pungent aroma; most of it is used in grated form. Often dusted on cheese sticks, garlic toast, spaghetti, etc.

pärmi — *Estonia* yeast.

pasas — *Sp* raisins.

paskalya cöregi — *Turkey* a lightly sweetened, yeast-leavened cake or loaf, an Easter specialty.

pasta — a product of the same type as macaroni, spaghetti, and noodles; also called alimentary pastes, macaroni products, etc.

paste — term generally applied to a "dead" (unleavened, non-elastic) coherent mixture; almond paste is an example.

pastej — *Sweden* pie or tart.

pastel — *Sp* cake, such as a layer cake, or pastry such as pie.

pasteurization — relatively mild heat treatment sufficient to kill vegetative forms of the predominant spoilage organisms in foodstuffs; not synonymous with sterilization.

pastie — (also, "pasty") a savory pie of the Welsh type, generally made in single-serving size. They usually have a non-flaky crust and, often, a beef-onion-potato filling; in shape like a fried pie, but baked.

pastillage — *Fr* gum paste used in making candies, decorations, etc.

pastisetas — *Mex* octagonal-shaped butter cookies.

pastry — an inexact term that is now generally regarded as applying to almost any dessert-type baked food.

pastry bag — a conical canvas (or plastic) bag that narrows to a small open point at one end and has a large opening at the other end; it is filled with a decorating paste (or deposit cookie dough), the open end is folded over and held closed, and hand pressure is used to force out the contents in a narrow strand from the pointed end; usually, a metal or plastic orifice is inserted in the pointed end to give some decorative treatment to the extruded material.

patata — *Sp* potato; *Mex* sweet potato.

pâte à chou — *Fr* cream puff paste.

pâte à pâté — *Fr* any of the various kinds of unleavened dough sheets used to cover, for example, meat pies. Often a simple lard, flour, salt, and water mixture.

pâte bâtarde — *Fr* standard formula dough for making ordinary white bread.

pâte feuilletée — *Fr* puff pastry dough.

patent flour — product made from the finer and whiter flour streams; comes mainly from ground purified middlings. Has lower bran content and higher endosperm content than straight flour, thus lower ash and lower fiber content.

pâte sablée — *Fr* rich sweetened shortcrust pastry, as used for certain sweet tart crusts.

pavlova — *Australia* a base of meringue filled or covered with a fruit preparation (many variations) and topped with whipped cream. Has been called the Australian national dessert.

pay — *Mex* pie.

peak — (1) Stage of optimum goodness. (2) In preparing meringues and some other whipped goods, it is the stage at which the mixture will form a persistent point when the beater is pulled vertically from the beaten mixture. A "dry peak" is short and stiff and has a rather dull appearance, a "wet peak" is long, soft, bending, and has a rather glossy appearance.

peaked — a cake or other baked product that rises up in the center to an excess degree, giving a somewhat pointed appearance.

peanuts — the ground nut, goober peas. The plant is actually a legume, but its seed is similar in many ways, including edible usage, to tree nuts. Common varieties are Virginia, Spanish, and Runner. Peanuts are roasted before use, except in one or two rare applications.

peanut butter — a paste produced by milling roasted (and usually blanched) peanuts to a fine particle size. Hydrogenated vegetable oil and emulsifiers are usually added to prevent oil separation and modify the texture. Salt is always added, except in dietetic products.

peanut oil — oil pressed from unroasted shelled peanuts. Used mostly for culinary purposes since it is, under normal market conditions, too expensive to compete with soybean or cottonseed oils for bakery use.

pecans — seeds of the tree *Carya illinoensis*. The nutmeats have a slight resemblance to small walnut meats, but their length to width ratio is greater. The flavor is mild and pleasant. A premium nut, primarily restricted to North American use. Does not need to be roasted before use in bakery products. Obligatory in pecan pies, of course.

pectin — a substance found in most fruits that causes gelling of sugar solutions under certain conditions, leading to the formation of jellies and jams. One of the "dietary fiber" substances, although it doesn't form fibers.

peel — wooden or metal paddle with a long handle used to place loaves (or pans) in an oven and take them out.

peel ovens — large stationary ovens, so called because the wide and deep baking hearths have to be loaded and unloaded with peels.

Pekar test — see "slick test."

peksimet — *Turkey* similar to bazlama (q.v.) except baked in a peel oven.

pelican — a spout sampler for obtaining a representative sample from a falling stream of grain.

pelle — *Fr* peel.

pekar test — same as "slick test."

pentosan — a polymer of pentoses, the latter being five-carbon sugars. The pentose analog of starch.

pepparkakar — *Sweden* gingerbread.

pera — *Sp* pear.

permeable — describes a film or sheet of a material through which molecules or ions can pass, usually selectively.

peroxide value (PV) — oxygen can react with a fatty acid chain to form peroxides or hydroperoxides. Peroxide value is a measure of the amount of these materials that have been formed in a sample. It is expressed as milliequivalents of peroxide-oxygen combined per kilogram of fat (meq/Kg).

petit bouchée — a miniature patty shell of puff pastry used with savory fillings to make a single-bite hors d'oeuvre.

petites galettes salées — *Fr* small salted crackers.

petit fours — according to current U.S.A. usage, small chemically leavened cakes, usually square, covered with colorful (and usually decorated) icings. The thin cake layers are often alternated with jam or other fillings. Frequently based on lean pound cake formulas or firm layer cake slices. Never treated as multiple serving pieces.

petit pan — *Fr* bread roll or small loaf.

pétrissage — *Fr* the kneading process.

pfeffernusse — pepper nuts, a Christmas cookie developed in Germany. In its usual versions, appears as a small ball-shaped cookie coated with powdered sugar. Texture crisp. Has a sharp spicy taste. Formulas usually include a small amount of black or white pepper.

pH — measure of the acidity or alkalinity of an aqueous system, pH 7 being neutral while lower values are acid and higher values are alkaline.

phosphated flour — wheat flour to which monocalcium phosphate has been added to serve as the acidic component of a leavening system, i.e., only soda has to be added by the user to cause the dough or batter to rise.

phospholipids — organic compounds with a lipid moiety attached to a phosphoric group, lecithin being an example. Many phospholipids have some emulsifying action.

phytase — enzyme that breaks down phytin into smaller molecules.

phytin — phosphorus compound found in wheat bran (and elsewhere) that is one form of dietary fiber; it binds some metallic ions.

pidesi — *Turkey* flatbreads in general.

pie — unleavened pastry shell containing a filling of sweetened fruit, pastry cream, custard, etc., and topped with another layer (or strips) of pastry or with whipped cream, meringue, streusel, etc. Thousands of variations are possible, including meat pies, fried pies, etc.

pie pin — a thin, relatively long rolling pin used for sheeting pie doughs.

pie press — a machine for automatically forming pie crusts out of lumps of dough by forcing the dough into a mold.

pie rimmer — a device for trimming the excess dough from the edge of a crust, sometimes provided with attachments to form decorative rims.

piernik — *Poland* cake.

GLOSSARY OF MILLING AND BAKING TERMS 83

pignolis — small white nuts taken from a tree of the pine family.

piima — *Estonia* milk.

pikelet — *UK* a product very similar to English muffins and crumpets, the differences not being clearcut or universally agreed upon.

pilaf — (various spellings) cooked rice (sometimes cracked wheat) with minor amounts of various other ingredients such as meat or vegetables

piña — *Sp* pineapple.

pinch — an inexact ingredient measurement, considered to be about one-eighth teaspoon.

pineapple — fruit of a tropical plant, grown in large amounts in Hawaii and Mexico. For the baker, canned pineapple (sliced into discs or wedges, or crushed) is the usual ingredient, but candied pineapple is often found in fruit cake.

pint — in the U.S.A., 16 fluid ounces; in the UK and Canada, an Imperial pint is 20 fl oz.

piparkoogitort — *Estonia* gingerbread tort.

pipe — to extrude cream puff paste, filling, icing, or piping jelly out of a piping bag, which is very similar (if not identical) in shape and method of use to a pastry bag.

pirukatäited — *Estonia* pastry filling.

piskótatészta — *Hungary* sponge cake.

pissaladière — *Fr* a type of flan said to be "first cousin to an Italian pizza;" a specialty of Nice.

pistachios — small to medium sized nuts, the meats being variegated greenish and light brown in color and having a distinctive flavor. The shells are often dyed red, but are normally ivory to white in color. Formerly they were obtained almost entirely from Iran, but superior quality nuts are now available from U.S. sources. Less subject than other nuts to rancidity.

pistolet — *Fr* split roll, something like a club roll.

pita bread — a thin, circular yeast-leavened bread that, due to the method of baking, has the top and bottom crusts substantially completely separated except at the edges. Used to form pockets that are filled with savory preparations.

pizza — a flat dough piece on which additives such as cheese, sausage slices, and tomato sauce can be placed; usually round, thin, and baked, but thousands of variations have appeared.

placek serowy — *Poland* cheese cake.

plait — a way of forming fancy breads and sweet goods; consists of intertwining strips or strings (usually three) of dough.

planetary mixers — vertical mixers in which the agitator revolves around its own axis while the axis describes a circle about midway between the center and the side of the bowl.

plansifter — a machine in which a number of oscillating sieves are arranged one above the other; used to separate the components of a mill stream.

plath — a Welsh type of flatbread.

plating — (1) Applying a very thin layer of dissimilar metal to a metal object. (2) Depositing a thin layer of flavoring or coloring material on to fairly large crystals, as of granulated sugar; usually accomplished by gentle mixing of the crystals with a concentrated solution of the additive, which is then allowed to dry. (3) Preparing a culture of microorganisms on, say, agar plates.

plaque — *Fr* a sheet of metal or other material on which bread or the like is baked.

plasticity — the consistency (texture or "feel") of a solid or semi-solid shortening.

plasticizing — a method of treating a shortening or other fatty-type of ingredient so as to develop a crystal structure that have improved processing response and organoleptic characteristics.

plastic range — the upper and lower temperatures between which a shortening has a suitable consistency for use in creaming and the like.

plátanos — *Sp* bananas, especially those of the large green type.

platform scales — weighing devices that have a large flat platform on which to place the material being weighed.

pliable — describes a dough that has a good consistency for processing, generally understood to mean soft but extensible.

plum pudding — a blend of raisins, currants, citrus peels, milk, suet, and spices, with just enough flour to bind these ingredients together; cooked by boiling or steaming, the pudding being held in a cloth. Many variations have been described, including baked types.

pneumatic equipment — devices relying on controlled pressurized air for their motive power.

pocket — in a divider, the cavity into which the dough is pressed for volumetric measurement.

pogácsa — *Hungary* biscuits in general; yeast-leavened or unleavened.

pogne — *Fr* rich bread, similar to brioche; sometimes filled with candied fruit.

pointage — *Fr* bulk fermentation.

polvore crescente — *It* baking powder.

polvorones — *Mex* a type of pan dulce, usually in the form of small, circular, moderately thin buns. May be flavored, as with chocolate.

polydextrose — a nonsweet polymer of glucose that is primarily useful as a bulking agent. Since the claimed energy yield is one calorie per gram, it could be a useful reduced-calorie substitute for starch.

polyester — usually refers to polyethylene terephthalate, a plastic resin much used for packaging applications requiring good strength and transparency. It has fairly good resistance to moisture vapor transmittance.

polyethylene — a plastic resin (polymer of ethylene) that has found very many applications in packaging, both as a film and for bottles. It comes in high and low density varieties, oriented and non-oriented, and with other modifications; also available as films combined with many other materials.

polyglycerol ester — emulsifiers produced by first polymerizing glycerol and then esterifying the polymer with selected fatty acids.

polymerization — (1) A chemical combining of two or more molecules of the same type, leading to long chains that may or may not be branched and that have qualities very different from the monomer. (2) A deterioration in frying fat resulting from the combining of fatty acids to give gum-like substances that cause foaming and other undesirable phenomena.

polymorphism — having the potential of existing in more than one crystalline form. Many fats can exist in at least three different crystalline forms — alpha, beta, and beta prime. The crystal form of a fat has an important effect on its melting point and its performance in use.

polypropylene — a plastic resin formed of polymerized propylene. Used in many of the same ways as polyethylene, but PP is generally tougher, harder, and (in some forms) more transparent than PE. Improved properties are found in oriented films, i.e., films that have been stretched in one or two directions before they have completely cooled.

polysaccharide — a polymer of several sugar molecules, such as starch (which is a polymer of glucose).

polysorbate 60 — commonly known as Tween, this emulsifier is used in cakes and adjuncts to give greater volume, and in bread to retard staling and strengthen doughs.

polystyrene — a thermoplastic resin consisting of polymerized styrene. It is used primarily in rigid and semi-rigid items rather than in films. Has good transparency and is cheap.

polyvinyl chloride — a plastic resin used mostly for making packaging films. Good grease and solvent resistance, but not particularly good moisture resistance.

polyvinylidene chloride — a plastic film tradenamed Saran. Very good strength and low moisture vapor transmission.

poolish — *Fr* a baking intermediate similar to sourdough starter.

poori — *Ind* a chapati that has been fried in deep fat.

poppy seeds — very small, kidney-shaped, bluish-black seeds with a nutty flavor. Used as whole seeds for scattering across the tops of bread rolls and in cakes, and (when ground) as fillings in kolache and other European pastries.

porgandi plaadipirukas — *Estonia* open-faced carrot pie, made with a yeast-leavened dough.

porosity — (1) Average size or size distribution of the cells in the crumb of a loaf of bread; described as open (large cells) or close (small cells). (2) The extent to which unusually large cells are found in the crumb of bakery foods; this kind of "porosity" is desirable in English muffins but not in most other bakery foods.

positive displacement pump — a pump that uses a reciprocating or rotating cavity to deliver a predictable volume of fluid at each complete cycle.

potable — suitable for drinking by human beings.

potassium bromate — a chemical substance ($KBrO_3$) used as an oxidizer in the baking industry. It is relatively slow acting.

potassium chloride — used as a substitute for salt (sodium chloride) in foods intended to appeal to those persons who believe the latter substance is deleterious to their health.

potassium iodate — a chemical used to oxidize dough proteins. It generally acts more rapidly than potassium bromate.

potassium sorbate — a preservative that acts similarly to sorbic acid but is often easier to use because of its greater solubility.

potato flour — a powdered material prepared by dehydrating cooked peeled potatoes.

pound cake — in modern terminology, this name is applied to a firm yellowish cake of fine cell structure, rather dense, and often in loaf form; the flavor is usually mild lemon or vanilla. The traditional formula is a pound each of flour, sugar, butter, and liquid whole eggs, but many different formulas are being used today.

powdered sugar — cane or beet sugar that has been very finely ground; it is available in different particle sizes. Commercial products contain about 3% corn starch and occasionally small amounts of other additives to prevent the formation of lumps due to moisture absorption. Synonyms: confectioners' sugar, icing sugar.

praline — a filling or flavoring for chocolates and the like that is made by roasting almonds (or other nuts) in molten sugar until they are brown and crisp, and then grinding the cooled "brittle."

pralines — filled chocolate candies in the European style.

precipitate — a solid material that separates from a solution as the consequence of some chemical or physical change.

preheat — allow an oven to reach, and stabilize at, the desired baking temperature preparatory to placing dough in the oven.

prepared flour mixes — blends of flours with other ingredients that are intended to reduce the measuring steps required at the point of mixing.

pre-scoring — cutting English muffins at their sides from edges to the middle to make it easier for the consumer to split the muffin into halves.

preservatives — materials added to foods for the specific purpose of increasing shelf-life; usually applied to chemicals that prevent or retard microbiological growth (e.g., sorbic acid).

press out — to divide a batch of dough into a specified number of pieces by means of a dough-press (bun press, roll press) machine.

pressure depositor — a device for extruding doughnut dough and the like from a pressurized chamber.

pretzels — formerly applied only to crisp snack items with a unique shape, formed from yeast-leavened dough and dipped in an alkali solution before baking. Now, also applied to soft products of the same general shape and to crisp products made in the form of sticks, nuggets, and fancy shapes.

product stream — any one of 125 to 150 mill streams in the flour manufacturing process.

profiteroles — little hollow balls baked from cream puff paste, often used as the basis for constructing multiple-serving desserts of elaborate design.

proof — expansion of the dough from the time it is formed into a loaf (or other final shape) until it enters the oven, or the period during which this occurs.

proof box — a chamber or room equipped with means for maintaining relatively constant temperature and humidity and for transporting dough containers such as loaf pans through the space.

proofing — the final rising of bread dough (except for oven spring) occurring after the dough has been formed into the final piece.

pro-oxidant — a substance that promotes or catalyzes an oxidation reaction. For example, materials that accelerate the development of oxidative rancidity in a foodstuff containing fat.

propionates — salts of propionic acid used as preservatives, they are primarily effective against mold growth.

propylene glycol esters — lipophilic emulsifiers used to improve whipping response and volume in cake batters and whipped toppings.

propyl gallate — an antioxidant.

protected active dry yeast — an ADY that contains antioxidants and is dried to a lower moisture than the regular form of ADY; as a result, it has a longer shelf life.

proteins — large molecules composed of amino acids; proteins are ubiquitous in living tissues.

proteinase — (protease) an enzyme that splits protein molecules into peptides or amino acids.

proving — same as proofing, the former spelling being more common in England.

PSIG — gas pressure (in pounds per square inch) measured without reference to atmospheric pressure.

psychrometer — a device for measuring relative humidity.

psychrophilic — describes microorganisms that can grow and reproduce at relatively low temperatures, e.g., 40°F.

puf böregi — *Turkey* chemically leavened puff pastries that are fried in shallow oil; generally made in small pieces; with various fillings, both savory and sweet.

pudding — (1) A soft sweet gelled product often texturized with a starchy material and flavored with almost anything. (2) *UK* Fairly lean doughs made up into multiple-serving pieces, say 1 lb or more, and cooked by boiling or steaming; though alleged to be a dessert-type product, the dough is not very sweet, but is enriched by a fatty material (traditionally suet) and raisins or the like; often served with a sauce of custard formulation and consistency, or jam.

puff paste — an unleavened dough that has been interlayered with butter or other shortening, then rolled out, folded over and sheeted again. The folding and sheeting operation may be repeated several times. If properly made, these doughs yield flaky, tender, and glossy baked products.

puff pastry margarine — a margarine specially made for layering puff pastry doughs. It is firmer than regular margarine because it contains fats of higher melting point; it usually contains very little salt and not much water.

puliszka — *Hungary* cornmeal dumplings.

pullman bread — bread baked in a pan of square cross-section having a cover. Sandwich bread.

pullman pans — long metal pans of almost square cross-section fitted with removable lids.

pumpernickel — in its original form, this is a very dense rye bread made entirely of crushed rye kernels with no admixture of wheat flour, and baked for a very long time (many hours) to develop a caramelized flavor. Now applied to many other kinds of dark rye bread.

punching — deflating, or pressing the excess gas out of a fermenting dough or sponge by kneading it. This is a process used in the manual preparation of doughs. In large operations, the remix step accomplishes the same effect.

pup loaf — a small (100 to 150 g) loaf of bread prepared in a laboratory procedure to determine the baking quality of flour or other ingredients.

puri — *India* chapati dough sheeted and deep-fried; small puffed breads fried in ghee.

purifier — a device in which an air current is forced up through a bed of middlings bouncing on an oscillating sieve so as to blow the bran out of the mixture and into a separate take-off tube.

pyaz — *Ind* onions.

pyrometer — a measuring device designed to determine high temperatures.

-Q-

qualitative — describes an evaluation or description that does not contain numerical measurements determined by actual testing.
quantitative — an evaluation or description that includes measurements, as of degree or amount.
quart — in the U.S.A., a quart is 32 fl oz; in the UK and Canada, an Imperial quart is 40 fl oz.
queso — *Sp* cheese.
quesadilla — *Sp* cheesecake.
quiches — many of these products are made in a crust very similar to a fruit pie crust, while others (especially in Europe) are placed on a yeast-leavened dough base. The filling is nearly always a non-sweet custard-like concoction including vegetables, meat pieces, etc.
quick breads — bread- or roll-like products such as muffins, biscuits, etc. They are almost always made from lean chemically-leavened doughs.
quick-cooking oats — oat groats that have been cut into several pieces and then rolled into thin flakes.

-R-

rack — a metal frame for holding trays or pans in vertical array; may be mobile or stationary.
rack cooler — a cooling chamber, usually provided with circulating refrigerated air, that has provisions for taking in and pushing out racks of baked products or of doughs to be retarded.
rack ovens — ovens in which racks (usually on casters) containing the pans are pushed in and pulled out of the baking chamber either mechanically or manually.
radiation — the transfer of heat energy by electromagnetic rays, such as infrared rays.
råg — *Sweden* rye.
rågbröt — *Sweden* rye bread.
raghif — *Israel* thick flat bread made from whole wheat flour and other ingredients. Traditionally baked on a bed of pebbles.
raisins — dried sweet grapes, may be dark or bleached.
ram — that part of a dough divider which forces dough into the compression chamber.
ramazan pidesi — *Turkey* a yeast-leavened flat bread similar to nan.
rancidity — a type of fat deterioration which causes unpleasant odors and/or tastes; can be of either the oxidative or the hydrolytic type.
rántott borsó — *Hungary* very small lumps of a dough made from egg, flour, milk, and salt, and fried in shallow oil; the dough is pressed through a

sieve or extruded in some other manner directly into hot lard. Have been called "fried soup peas."

rapeseed — the grain from which canola oil is pressed; the small, round, dark-colored seeds have been used as the measuring medium in volumetric displacement devices designed for determining the specific volume of baked products.

rarou — *Egypt* a thin crisp bread that is consumed after moistening or is crumbled like crackers and eaten in soup, etc. The dough consists of a mixture of equal proportions of wheat flour and corn flour, 10% okra flour, salt, and water. After the dough ferments for about an hour, it is pressed into a thin layer and baked.

ratio — the quantity of two or more materials as related to one another.

ravioli — two small pasta sheets completely surrounding a filling of meat, cheese, etc. Usually boiled and served with a sauce, as of tomato.

raw materials — essentially all the materials used to make up a product; ingredients, even though some of them may be cooked, not "raw." The preferred term is "ingredients."

rebanadas — *Mex* sliced and toasted white bread spread with butter-flavored cream and powdered sugar.

recipe — a list of the ingredients and their quantity as needed for preparing a product; differs from "formula", if at all, in that recipes are denominated in household measurements and yield small quantities of product while formulas are in metric amounts and are used to prepare wholesale quantities (or lab quantities) of product. Most recipes also include processing instructions while formulas generally do not, though there is no hard and fast rule.

reciprocal baking — a method of allocating production capacity among a multi-plant bakery so that one plant will bake one type of product and another will bake a different type; the plants then ship items through regional centers for final assembly, distribution, and sale to customers.

reciprocating slicer — equipment for slicing bread loaves that moves in an up-and-down motion saw-like blades set in a frame.

reconstitute — to rehydrate a dried material to more-or-less its original water content; sometimes, less accurately, used to mean thawing a frozen product.

recovery periods — the rest stages that allow dough pieces to accumulate gas and the gluten to relax following, for example, rounding.

red dog — not a carmine canine, but a very low-grade flour with a high content of bran. This by-product is not often used for bakery products, but with the current craze for high fiber products it could become a premium ingredient.

reduced iron — very small particles of metallic iron; has been used as a mineral supplement in bakery products and some other foods.

reducing agent — an ingredient such as glutathione or L-cysteine, that adds hydrogen to the disulfide bonds in gluten, resulting in a weakening of the protein network.

reducing sugar — a sugar that reduces Fehling's solution; related to the position of the aldehyde or ketone group in the molecule. Most sugars are reducing sugars, but sucrose is not.

reduction — procedure by which grain is milled into flour and by-products.

reel — a rotating, generally cylindrical, sieve, without beaters inside and in most cases horizontally aligned.

reel ovens — early name for the revolving tray oven; now used mostly for the smaller type of revolving tray ovens, as found in some retail bakeries.

refractive index (RI) — the refractive index of a substance is the numerical expression of the ratio of the speed of light in a vacuum to the speed of light in a test object. For practical measurements, the scales of instruments show the refractive indexes compared to air rather than vacuum. Simple, relatively inexpensive instruments utilizing this principle are often used to measure the so-called soluble solids in certain kinds of foodstuffs, e.g., sugar syrups, corn syrups, and fruit juices, and to identify and characterize fats and oils.

refrigerants — (1) Materials, such as liquid nitrogen, which can be used to cool products. (2) Heat transfer media used in refrigeration systems, including freezers.

relative humidity — the ratio between the amount of water vapor actually in a sample of air, and the greatest amount of water vapor the same air could contain, both being at the same temperature and pressure.

release — in flour milling, the amount or percentage of stock yielded by a break step and which does not go on to the next break step, but is sent to purifiers, etc., for grading.

remonce — *Denmark* pastry filling, as butter cream.

rendering — application of heat to a fatty animal tissue in order to release the fat for collection or further processing.

residual sugars — sugars that remain in a finished bakery product, i.e., sugars that have not been fermented.

response — the reaction of a dough to the set of conditions it encounters during preparation.

rest period — the time given a dough to recover desired properties after some processing step has been applied.

retarder — a refrigerator used for retarding doughs.

retarding — delaying fermentation of dough by chilling (not freezing) it.

rétes — *Hungary* strudel.

rétestészta — *Hungary* strudel dough.

retrogradation — change of a cooked starch from a gel to an insoluble state, said to be due to formation of aggregates of the starch molecules.

returns — (1) Mill stock that is sent back for retreatment to the same machine or even to a machine nearer the head of the mill. (2) Finished products that have been sent back to the manufacturer by a wholesaler, retailer, or consumer because the products are outdated or have other unacceptable characteristics.

reversion — an undesirable change in the flavor of a refined oil or fat, so-called because it was thought to result when a refined, nearly flavorless oil reverted to its original "raw" flavor.

revolving oven — a reel oven.

rheology — the science concerned with the deformation and flow of matter.

riboflavin — a B-vitamin required to be added to enriched flour.

rice — the grain obtained from the plant *Oryza sativa*. Has a pleasant mellow flavor, but contains no gluten so that an extensible dough cannot be made from it. Commercially available forms are rough rice (with the hulls on), brown rice (not polished, but hulls removed), and polished rice (the usual form of commercial rice). Wild rice is an entirely different grain.

rice flour — finely ground rice kernels; used as an ingredient and as dusting powder.

rice paper — an edible paper-like material made of a baked rice starch batter; it is used as the base for very sticky confections that would otherwise be impossible to handle.

rice starch — starch separated from rice grains; it has slightly different properties than corn starch. This term is often applied to rice powder, which is actually finely ground white rice containing other substances (such as proteins) in addition to starch.

rich doughs — doughs that contain considerable amounts of some or all of the enriching ingredients, such as shortening, milk, sugar, etc.

ripe dough — a dough that has reached a stage of fermentation and conditioning making it suitable for make-up.

ripeness — readiness of dough for baking.

ris — *Sweden* rice.

rise — the increase in height (or, less often, in volume) attained by a piece of dough in a given fermentation time.

riso — *It* rice.

rocks — small, rough-surfaced fruited cookies baked from a stiff batter.

rock sugar — very large crystals of sucrose.

rognures — *Fr* scraps of leftover dough.

roles glass — *Mex* sweet rolls, topped with sugar glaze and containing a filling of cinnamon and/or raisins.

roller system — the modern method of reducing wheat to flour, based on pairs of rotating cylinders between which grain and partially milled material are passed; a roller mill will also include many other devices, e.g., purifiers, sifters, cleaners, and dust collectors.

GLOSSARY OF MILLING AND BAKING TERMS 93

rolling pins — wooden or metal cylinders, 1 to about 4 inches in diameter and a foot long (more or less), with a handle at each end, used for manually sheeting out dough.

rolls — portion-size pieces baked from lean yeast-leavened dough; the individual serving version of bread loaves; buns.

rope — a type of spoilage occurring in bread, now rare. Characterized by a stringy, gummy degeneration of the crumb. Caused by *Bacillus mesentericus*.

rosca de panque — *Mex* ring-shaped cake topped with powdered sugar.

rosette — (1) A bud-like ornament made from icing and somewhat resembling a rose. (2) A cookie or other flat disc with scalloped edges.

rosquilla — *Sp* doughnut.

rotary cookie — a cookie molded on a rotary or Dutch machine, the latter devices having a large metal roller with molding cavities cut into its surface.

rotary oven — an oven having a horizontal rotating hearth that is circular in shape, now largely obsolete. The term is sometimes incorrectly used for revolving tray ovens.

roti — *Ind* bread.

roti flour — *Ind* a high extraction wheat flour of rather coarse granulation, widely used in Asia to make many types of unleavened bread, such as paratha and roti.

roughage — non-digestible material in food; approximately the same as dietary fiber.

roulade — a layer of baked cake that has been spread with jam or some other kind of filling then rolled up; a jelly roll is an example.

rounder — a machine that forms balls from roughly cut (scaled) dough pieces.

rounding — rolling cut dough pieces coming from the divider so as to seal the surfaces and form a uniform skin on the pieces.

round top bread — open top bread; loaves baked in pans that have no lids.

rout press — a bar press machine for extruding strips of cookie dough.

royal icing — decorative frosting of cooked sugar and egg whites.

rukkijahu — *Estonia* rye flour.

rum — an alcoholic beverage distilled from fermented sugar cane juice. Often used as a flavor in bakery dessert products (babas au rhum, etc.) and in confections.

rundstykker — *Denmark* bread rolls in oblong form.

rye — a grain that is darker and somewhat more fibrous than wheat and yields a flour that is stronger flavored but is much less effective in contributing structure to bread than is wheat flour.

rye bread — bread made from a blend of rye flour and wheat flour; rarely, made from all rye flour, resulting in a dense, moist bread.

-S-

saccharify — convert into sugar.

saccharin — a chemical substance about 200 to 300 times as sweet as sugar; some consumers find it has a bitter flavor, as well.

saccharine — having the flavor nature of sugar, sweet.

Saccharomyces cerevisiae — the genus and species names of bakers' yeast.

sack cleaner — a device for removing flour and other dust from the outside of sacked ingredients, either immediately after the sack is filled or just before the sack is dumped.

Sacher torte — a very rich confection consisting, in its traditional form, of a layer of baked chocolate cake covered with apricot jam and enrobed with chocolate icing.

safflower oil — a food oil of high unsaturation pressed from safflower seed.

saffron — a spice consisting of the dried stigmas of the flower *Crocus sativus*. Used in saffron buns, and the like, to give an orange-yellow color and a unique aromatic/pungent flavor. Very expensive.

saga-bolo — *Jap* a Japanese confection made from a chemically-leavened, very sweet dough. Sometimes, a mixture of buckwheat and wheat flours are used in this product.

saggina — *It* buckwheat.

sajtos izelitö — *Hungary* hot cheese crackers; a sort of savory sandwich cookie. The base cakes are disks of a baked unleavened dough of flour, butter, Swiss cheese, sour cream, and paprika; a paste of butter, cheese, and sour cream is placed between two baked discs.

saka-manju — *Jap* steamed Japanese bun made from fermented dough and containing bean jam filling.

sal — *Sp* salt.

salad oil — a refined, bleached, and deodorized edible oil that has probably also been winterized. Some oils are natural salad oils, i.e., they do not form fat crystals at refrigerator temperatures.

Sally Lunn — a very old type of rich bread in muffin form that originated in England. Many formulas are extant.

salt — unless the word is modified, it means sodium chloride, which is sold in a wide range of purities and particle sizes.

saltstaenger — *Denmark* salt finger rolls, salt sticks.

salt-rising bread — a type of sourdough bread for which the dough is ripened partly by a kind of salt-tolerant yeast and partly by bacteria.

samosa — *Ind* stuffed meat pasties.

sanding — applying coarse sugar crystals to the surface of confections or bakery products.

sandkuchen — *Ger* a baked product of the pound cake type, but with about half the wheat flour replaced by wheat starch.

sandwich cookie — a confection consisting of two baked cookie wafers with a creamy filling in between.

sandwiching machine — a device for combining base cakes and filling to make sandwich cookies.

sangak — *Iran* sourdough flat bread made from flour water, salt, and starter; sometimes sprinkled with sesame or poppy seeds. Traditionally baked on a hearth covered with pebbles.

sanitation — the methods and practices used to prevent contact of foodstuffs with environmental contaminants.

santara — *Ind* orange (the fruit).

saponification (Sap) value — one of the tests performed to characterize edible fats and oils. Saponification is the hydrolysis of glycerides (such as fats) with a caustic or alkali to form free glycerol and fatty acids in the form of soaps. The reaction mixture is titrated with standardized alkali to get the saponification value, which is inversely related to the average molecular weight of the fat and, is therefore, an indication of the type of fatty acids in the fat. The larger the fatty acid molecule, the lower the SAP value.

sarazin — *Fr* buckwheat.

saturated steam — steam containing the maximum amount of water that can be held as vapor at that particular temperature and pressure.

sausage roll — puff pastry rolled around a cooked, ground meat mixture containing spices, then baked.

scale — (1) Any kind of weighing device. (2) To cut dough into pieces of the desired size (by weight or volume).

scaling — (1) Cutting or otherwise separating a mass of dough into smaller pieces prior to further processing. The pieces will be uniform in weight but not necessarily of the same weight as the finished item, since some products require the assembling of one or more pieces to yield the shape and size deemed desirable. And, of course, the dough piece can decrease in weight due to loss of moisture and other volatiles, and may gain in weight due to the acquisition of dusting flour, adjuncts, etc. (2) Weighing out ingredients.

scalp — to sift, as when separating the coarse part of the mill grind. Also used to mean the removal of foreign material from the grain as it was being cleaned before milling.

Schaal test — the Schaal oven method for determining the oxidative stability of a fat-containing food involves heating a given amount of sample in a covered glass container at an elevated temperature until a rancid odor is detected. Results are reported as the length of time elapsing until the end point is reached. Temperatures, sample size, container type, etc., vary from analyst to analyst and should be specified when test results are recorded.

This test can be quite useful if properly conducted and interpreted.

schedule — or, shop schedule. A sheet or form used in the bakeshop to record the types and amounts of products to be made and the sequence and timing of their manufacture.

schwartzbrot — *Ger* a very dark, dense, moist type of rye bread; a name that is sometimes applied to types of pumpernickel.

scone — a class or type of individual-serving hot breads that includes many varieties. Scones originated in Scotland as oatmeal-based cakes, but now virtually all of them are made of wheat flour. They are used in very much the same way as the baking powder biscuits popular in the southern U.S., but scones are usually a little richer and may include raisins. Some varieties are cooked on a griddle.

scoop — a shovel or ladle with rounded end and short handle, sometimes calibrated for volume; used for transferring dry ingredients to a scale pan or mixer bowl.

scorch — to overcook the surface of a product (especially with an open flame) so that the crust darkens excessively.

scoring — (1) Judging finished goods according to descriptive scales of quality factors. (2) Cutting or slashing the top surfaces of dough pieces.

Scotch bun — see "black bun."

scrap — product, or intermediate such as dough, which is unusable in its existing form because of some defect and must be reworked or discarded.

scratch — to form a dough, batter, etc., by weighing and combining all the individual ingredients in the bakery, as contrasted with the use of pre-mixes.

scratch rolls — in English usage, finely fluted mill rolls whose function is to remove bran fractions from the floury middlings.

screens — (1) A kind of baking container for hearth breads consisting of a wire-mesh form mounted in a supporting frame; usually made in the shape of a half-cylinder. (2) Flat wire-mesh rectangular supports on which doughnuts are proofed and then placed in the fryer. (3) Sieves for sifting.

scroll — a flake buster, q.v.

seam — the outer end of the dough sheet used to form a loaf; visible in the unbaked loaf as a slight indentation bordered by a ridge down the side of the dough cylinder.

seasoning — the addition of spices, salt, etc. to foods. Also, these flavors.

seb — *Ind* apple.

sedimentation test — a test for evaluating wheat protein quality; flour is suspended in an aqueous solution of lactic acid and held for a time under specified conditions. The volume occupied by the sediment is measured.

sembei — or o-sembei *Jap* rice wafers flavored with soy sauce or other seasonings. Rice and/or wheat flour is kneaded to make a dough, mixed with egg yolk, seaweed, spices, etc., then baked in a thin metal mold.

semolina — (1) Large middlings, passing through screens of 18 to 40 meshes per inch. (2) The purified middlings of durum wheat, used to manufacture the better quality of pasta.

senbei — *Jap* the generic name for Japanese crackers.

senesen — *Egypt* fiti (q.v.) made from a batter containing sorghum flour instead of wheat flour.

sepikujahu — *Estonia* wheat flour.

sesame seeds — small flat white seeds of the sesame plant; used for topping bread rolls. A cooking oil is also obtained from these seeds but it is not much used in Western cooking.

shamsy — *Egypt* rectangular- or disk-shaped bread with a light brown crust and a firm white crumb, made from a flour-water-salt dough. After the dough is formed into pieces, it is allowed to sit in strong sunlight for about three hours.

shamy — or "shamey." Yeast-leavened Egyptian flat bread made from 72% extraction flour. Baked so as to yield a "pocket," as in pita bread.

sharps — middlings.

sharp-to-sharp — the operation of roller mills when they are run so that the shorter face of the corrugated cutting edge on the faster rotating roller meets the longer face of the corrugations on the slower rotating roller.

sheen — the reflection or glossiness of the cell walls on a cut surface of the loaf; more generally, a shininess or matte gloss on the surface of any material.

sheet pan — a flat pan, often about 18x26 inches, with edges raised about an inch. Used for baking cookies, layer cakes, buns, hearth breads, etc.

shelf life — the length of time a product retains acceptable quality under average conditions of distribution and display.

shell top — in bread, a condition where the top crust separates from the loaf, giving the appearance of a cap over the loaf.

shingara — *Ind* vegetable pasties.

shipstuff — an old name for low-grade flour of high bran content.

shoofly pie — a brown sugar and molasses flavored cake that is baked in a pie crust.

shortbread — (1) Cookie made from butter, flour, and sugar with little or no liquid added. (2) Baking powder biscuit dough enriched with sugar and shortening, used as a basis for strawberry shortcake and the like.

short break — the condition that exists when a loaf exhibits minimal oven spring. The break and shred along the side is very narrow in such loaves.

shortening — a fatty ingredient used for the purpose of imparting desired textural qualities (especially, shortness, tenderness, or flakiness) to baked foods; sometimes applied with doubtful appropriateness to frying fats and oils. Shortenings may range in composition from pure natural oils to sophisticated mixtures of modified fats and additives.

short paste — a mixture of flour, fat, sugar, and water in various proportions. Generally much less elastic than bread dough.

short patent — a patent flour containing a relatively small percentage of the best flour streams; would not ordinarily be used to describe patent flour containing more than about 80% of the total flour output.

shorts — a low-grade mill product, containing principally germ and fine bran particles; used for animal feed.

short system — a milling process involving a relatively small number of reduction processes; it will yield comparatively few products.

short-time dough — a straight dough to which a higher than normal proportion of yeast has been added to make the dough ferment faster.

shred — appearance of the surface within the break area. It is described as smooth, ragged, broken, etc.

shrink film — a plastic film that can be draped around a package or product and then shrunk by applying heat, so as to get a tight fit.

shrink tunnel — the equipment used to finish the shrink film packaging process.

sieve — a device for separating particles according to size; consisting of a frame or vessel having near its bottom a screen with uniform openings.

sifter — an automatic sieving device, and particularly one in which horizontal (rather than cylindrical) sieves are used.

silicone-coated paper — paper that has received a coating of silicone anti-stick compound, often used as a pan liner.

silicones — compounds sprayed on surfaces (particularly the inside of bread pans) to reduce sticking of product.

silo — a very large storage bin, usually assuming the form of a cylinder on end; in some cases, other shapes are used. Can be used for both dry and wet ingredients.

simmer — to cook just below the boiling point so that just a few bubbles constantly form in the lower part of the vessel.

simnel cake — a special cake formerly made for the Easter season; it resembles an English meat pie in shape but its composition is more like that of a rich plum pudding. Usually has a strip of marzipan or almond paste on the top.

simple syrup — two parts sugar mixed with one part water, heated and stirred until all the sugar has dissolved. A stock intermediate having a number of uses in the confectionery, baking, and food service industries.

single-acting baking powder — a leavening system that generates carbon dioxide gas continuously when water is added; if not made up and baked promptly, products containing this type of baking powder may exhibit insufficient rise.

single lap oven — a traveling tray oven in which the baking shelves travel one long horizontal run during baking.

GLOSSARY OF MILLING AND BAKING TERMS 99

sinn — *Egypt* flatbread made from a dough consisting mostly of bran. It is yeast-leavened and given one or more fermentation steps.
siro — *Mex* sirup.
sizing — breaking down and grading the coarser middlings.
skeppsskorpor — *Sweden* hardtack.
skim milk — milk from which nearly all the butterfat has been removed by centrifugal separators.
skorpor — *Sweden* biscuits.
slabs — marble or slate table tops used by confectioners as a surface on that to work boiled sugar and other hot mixtures.
slack dough — a weak dough, having insufficient elasticity to process properly.
slick test — pressing and smoothing out a sample of flour and comparing its surface appearance to that of a standard sample, so as to qualitatively estimate the color, speckiness, etc. After the dry surface has been viewed, the sample is sometimes dipped in water to bring out additional details.
slow dough — a dough that ferments and conditions slowly, usually due to a temperature that is too low, too much salt, insufficient or stale yeast, or too much sugar.
slurry — a fairly concentrated aqueous mixture, but one sufficiently fluid to be handled by liquid systems.
smead — see "smid."
smid — *ME* finely ground semolina, often used as the basis for cakes and filled cookies in Middle Eastern cookery.
smoke point — the lowest temperature at which a fat sample, heated under a prescribed set of conditions, gives off a thin continuous stream of smoke. The smoke point of a frying fat should be as high as possible. It is typically dependent on the amount of free fatty acids and monoglycerides present in the oil.
smör — *Sweden* butter.
smörbakelse — *Sweden* pastry.
smorkage — *Denmark* coffee cake.
smutter — formerly, a general name for any wheat-scouring machine but now confined to a machine specifically designed to rid grain of smut, which is a type of fungal growth.
snack cake — a portion-sized cake, usually iced and filled, intended to be consumed without eating utensils as a between-meals dessert, essentially filling the same role as candy bars.
snack crackers — savory flavored crackers, often in unusual shapes, and usually in sizes smaller than saltines. May be made up similarly to sandwich cookies, i.e., two crackers enclosing a filling such as peanut butter.
snaps — small cookies that spread in the oven to give a thin layer which bakes into a crisp textured piece.

soba — *Jap* noodles made from buckwheat flour.
socker — *Sweden* sugar.
soda — baking soda is sodium bicarbonate; washing soda is sodium carbonate.
soda cracker — (saltine) a thin crisp cracker made from a yeast-leavened dough with a late addition of baking soda; the dough is lean in formulation, has been laminated just before cutting, and contains a large amount of a sponge that has been fermented up to 16 hrs.
sodium acid pyrophosphate — a leavening acid causing a rather slow release of carbon dioxide when combined with bicarbonate.
sodium aluminum phosphate — a common leavening acid; slow acting.
sodium aluminum sulfate — another leavening acid; slowest acting of all the common ones.
sodium benzoate — a preservative effective against many microorganisms when used in foods having a pH below 4.5.
sodium bicarbonate — baking soda.
sodium bisulfite — functions as a dough conditioner to weaken or soften dough; it was added to most cracker doughs at one time, but now is out of favor.
sodium diacetate — a preservative described as a "molecular combination of sodium acetate and acetic acid." Its preservative effect is probably due to the release of free acetic acid.
sodium erythorbate — an antioxidant.
sodium monoglutamate — a flavor enhancer that has very little flavor of its own.
sodium propionate — a preservative added to bakery foods to delay mold growth; has basically the same effect as propionic acid but is easier to use.
sodium steroyl lactylate — an emulsifier and dough conditioner with significant effects on processing response and finished product quality of many yeast-leavened and chemically leavened products.
soft ball stage — a stage in sugar boiling reached at about 245°F.
soft red winter wheat — like other soft wheats (*Triticum aestivum* varieties), the flour from SRW wheat usually contains less protein and a lower quality protein than flour made from hard wheat. The flours from soft wheat do, however, cover a wide range of quality, and some are even used in family or all-purpose flour. Generally, SRWW flours are best for cakes and other pastries, cookies, snacks, etc.
soft water — water with a comparatively low content of minerals, especially calcium. Distilled water is the ultimate in "soft" water.
soft wheat flour — see the comments under Soft Red Winter Wheat.
solid fat content (SFC) — the actual percent of solid fat in a sample of fat or oil, determined at standard temperature check points by pulsed nuclear magnetic resonance procedures. The analytical method is based on the fact

that protons in liquid fat are more mobile than those in solid fat. These more mobile protons are responsive to a magnetic field. By using this technique, the proportion of liquid fat which is present can be directly measured. Contrast with solid fat index.

solid fat index (SFI) — an empirical measure of the solid fat content of a sample of oil conducted at standardized temperature check points. It is based on a dilatometric procedure relying on volumetric changes occurring during melting or crystallization. SFI specifications are important in determining the behavior of a fat during processing, etc.

solid heat — if all the oven components have been brought up to the temperature they will maintain during continuous production, they are at solid heat. The term indicates that the temperature can be maintained within the desired range even when product is going through the oven at the top rate.

solution — a liquid in which another substance has been completely dispersed in ionic or molecular form. If molecular aggregates exist, they are not dissolved.

somen — *Jap* thin noodles (vermicelli) made from wheat flour.

sooji — *Ind* semolina.

soola — *Estonia* salt.

sopapillas — *Sp* chemically leavened small deep-fried cakes, somewhat like a doughnut without the hole; usually only slightly sweet.

sorbates — see sorbic acid, of which sorbates are the salt form. Sorbates generally dissolve more readily than sorbic acid.

sorbic acid — a preservative effective against many molds, yeasts, and bacteria. Depresses activity of baker's yeast and so is not added to yeast-leavened doughs. Formulators should check the flavor consequences of using it, especially if they expect the product to be heated by the consumer.

sorbitan esters — emulsifiers made by reacting sorbitol esters with fatty acids.

sorbitol — a sugar alcohol used as a humectant, or water binder, in confections and bakery foods.

sorghum — a cereal grain mostly used as animal feed, but also as a human food in some parts of the world. A different variety of the plant is used as a source of juice (pressed from the canes) which is evaporated to form sorghum syrup, a specialty flavoring sweetener with limited use in the Southern U.S.A.

sourdough — a starter or leaven which contains some basic dough ingredients plus a large population of yeasts and bacteria. Doughs made with these inoculants characteristically develop a sour flavor and pungent aroma. Many, perhaps most, sourdoughs contain large amounts of lactic acid bacteria.

sours — ingredients usually containing lactic and acetic acids as well as other flavoring materials that have been developed by yeast and bacteria growing over a period of time; used as flavoring and leavening ingredients, particularly in rye breads.

soybeans — a legume that produces small round seeds containing fairly high amounts of protein and oil. The beans are generally pressed and extracted to remove the oil, and the cake that remains is converted to animal feed or further processed to yield protein supplements and other food ingredients.

spaetzle — also, spätzle. *Ger* Little lumps of noodle-like dough usually formed by forcing the dough through a collander or the like, and cooked in soup much as noodles are cooked.

Span — a trade name for sorbitan monostearate, an emulsifier that has found some applications in chemically leavened foods.

spatula — a utensil shaped like a knife with (usually) a rounded end, but not sharpened; flat, thin, somewhat flexible, and often with a plastic or wooden handle.

specific gravity — the ratio of the weight of a given volume of sample material at 25°C to the weight of the same volume of water at the same temperature.

specific heat — the ratio of the quantity of heat required to raise the temperature of a body one degree to that required to raise an equal mass of water one degree.

specific volume — the reciprocal of specific gravity. An important factor in bread quality evaluations, where the figure is expressed as the number of cubic centimeters occupied by one gram of substance. Other units can also be used.

spekulatius — a German or Dutch butter cookie, molded in wooden forms, usually appears as rectangles with human or animal figures on top.

spice — any one of large number of materials derived from plants that contain odoriferous constituents suitable for use as food ingredients.

spiral cooler — equipment for chilling or freezing foods that carries the product along a continuous conveyor wrapped in a helical pattern about a central cooling duct.

spiral mixer — mixer with an agitator having an approximately spiral configuration.

sponge-and-dough — a method of producing bread in which a sponge is made by mixing a part of the flour with part or all of the water, and, usually, all of the yeast and yeast food; this mixture is allowed to ferment until it is judged to be ready for incorporation with the other ingredients to make the "dough."

sponge cake — made by combining a whipped mixture of whole eggs or egg yolks with the other ingredients. A pound cake is a type of sponge cake.

sponge flour — a flour for use in preparing bread sponges; may be the same kind of flour as used for the dough, or a stronger flour.

spoon bread — a type of corn bread that has a high moisture content and is cooked in a casserole so that it has a very soft consistency, in some cases almost like pudding.

spore — one of the reproductive forms of certain fungi, bacteria, and yeast; microscopic, usually spheroidal, dormant forms that germinate when provided with suitable growing conditions. Much harder to destroy than the vegetative form of the microorganism. Mold spores are found everywhere in the environment.

spotty heat — the condition existing when an oven has a very uneven heat distribution.

spray drying — the process of atomizing a fluid food product and delivering the mist to a chamber through which hot air is flowing. The resultant low-moisture powder has minimum heat damage and is usually in the form of a fine powder. Has been applied to many foods, such as eggs, milk, yeast, and cheese, to form storage-stable bakery ingredients.

sprayed cracker — the type of savory cracker (rich saltine type) that is sprayed with vegetable oil after it is baked so as to improve texture and appearance; often flavor is also improved.

spread — an important specification applied to cookie flours; it is based on the change in width of the test dough piece as it bakes.

spring — see, "oven spring."

springerle — Bavarian-type anise-flavored sugar cookies, formed by pressing dough into wooden molds, then air-drying the pieces before baking.

spring wheat — wheat (*Triticum aestivum*) that is sowed in the spring and harvested in the fall; generally has higher protein content and yields a stronger flour than winter wheats.

spritz — a method of forming cookie doughs by extruding a fairly soft batter through a tube so that some surface detail is retained from the orifice outline; also applied to similar extrusion processes used for other food materials.

sprouted wheat — fully matured wheat kernels that have been wetted while still on the plant, and have sprouted before being harvested. The quality of flour milled from such wheat is very bad. Some health food enthusiasts deliberately sprout wheat grain to make a bakery ingredient thought to have some dietary advantages.

spun sugar — sugar syrup that has been cooked to about 300°F and spun into threads by a mechanical process or by dipping a bundle of wires into the molten mass and waving it in the air. Used only for decorations. Cotton candy is made by a variation of this process.

stability — the relative resistance of a product to undesirable change. For fats and oils, stability may refer to resistance to oxidation, hydrolysis, flavor reversion, or development of off-flavors and off-odors. For doughs, the term generally refers to the range of fermentation time through which an acceptable product can be obtained. For many food products, such as bread, stability includes the resistance to deleterious microbiological processes.

stabilizer — any ingredient or additive that causes a food product to retain some desired characteristic for a longer period of time, as by delaying the collapse of foams or the breaking of emulsions. Bread stabilizers may have the function of delaying "staling," i.e., development of undesirable firmness.

stack ovens — deck ovens that can be stacked one on top of another for saving space.

staksill — *Sweden* semolina.

stale — a finished food product that has undergone physical and chemical changes rendering the product unacceptable from an organoleptic standpoint. Does not necessarily mean the product is harmful or lacks nutritional value.

staling — decrease in acceptability due to changes occurring after production; the changes may be due to chemical reactions or physical modifications, but "staling" is not generally regarded as including obvious mold growth (spoilage) or mechanical damage due to handling abuse (breakage).

Standard Deviation — a statistical term indicating the square root of the mean square of the deviations from the mean of a population of samples. Also called "sigma."

starch — a polymer of glucose found in many kinds of plants. Digestible by humans and forms a large part of the caloric content of many foods. When offered as an item of commerce, the ingredient is generally a fine white powder, dispersible but not soluble in water.

starch damage — physical changes (breakage, abrasion) of starch granules, usually an inadvertent result of milling or some other process. Such granules are much more susceptible to attack by amylases, and the amount of starch damage has been shown to be a factor affecting dough and bread quality.

steam — vapor emitted from water at its boiling point; confined water vapor held at a temperature above the boiling point at the pressure existing in the enclosure.

steaming — (1) Injecting steam into an oven while products are being baked. (2) Cooking products by holding them in an atmosphere of hot water vapor — applied to some dough products such as certain kinds of puddings, brown bread, etc.

steam kettle — versatile cooking equipment consisting of an inner vessel surrounded by a jacket (or pipes) through which steam is circulated. This equipment is made in a wide range of sizes and is found in many food plants including bakeries.

steam tables — omnipresent in cafeterias and most other restaurants, but also used in bakeries to keep icing, etc., warm. They consist of a table with openings into which containers of food can be put; below the top deck is a chamber into which low pressure steam is injected to keep a more-or-less constant (below boiling) temperature.

steam-tube ovens — ovens in which the baking chamber is heated by pipes containing steam under high pressure.

stearine — generally, a hard fat. Can be an oil that has been highly hydrogenated or the high-melting fraction from a rendering process, as in beef tallow rendering, where the other fraction is oleo oil.

steep — to put some solid substance in water for an extended period of time (hours) so as to cause the particles to become thoroughly equilibrated with the solution.

sterilize — reduce the microbiological content of a material (or space) to essentially zero by heating, irradiating, adding chemicals, etc.

sticky doughs — doughs that have excessive surface adhesion, making them difficult to process. Using too much water in the formula is a common cause, but an undue amount of punishment in the divider (or other make-up equipment) will have similar effect, as will excessive mixing, under- or over-fermentation, getting the dough too hot, and several other processing or formulation errors.

stiff dough — a tough, rubbery, bucky dough.

stollen — a moderately flat multi-serving loaf made from a rich yeast-leavened dough containing pieces of candied fruit, raisins, nuts, etc., usually lightly spice flavored. Often iced after baking.

storage life — the time between completion of processing and the development of unacceptable staleness.

storage temperature — the temperature (usually average temperature, sometimes a range) at which a product or ingredient has been, or should be, stored.

straight dough process — a method of bread-making in which substantially all the ingredients are mixed together at one time and then fermented, as contrasted with the sponge-and-dough process.

straight flour — all of the mill grind that can be designated as flour, what remains is millfeed. It can be divided into patents and clears, q.v.

straps — two or more loaf pans bound together by metal "straps" so they can be handled as one unit in the breadmaking process.

stratify — to spontaneously separate into layers, as when emulsions break down into an aqueous layer, an oily layer, and a solid layer. Making layers of dough intentionally is called "laminating."

strength — a rather inexact term applied to the baking quality of flour; it summarizes, in part, the contribution of the flour to extensibility, absorption, stability, and other characteristics features of the dough and baked product. Depends to a large extent upon the percentage of gluten in the flour, although many other factors can come into play.

strengthener — a wheat that is added to another wheat in a mill mix, to improve the quality or quantity of its protein content.

streusel — a crumb-like mixture of indefinite composition but almost always containing at least sugar, flour, and fat. Used for strewing on the top of the fruit in pies and coffee cakes, on crumb cakes, etc. Other kinds of mixtures have also been called streusel.

string icer — a machine that deposits thin strings or strips of warm icing on the top of sweet rolls, coffee cakes, etc. The more advanced equipment can automatically move the nozzles to give a simple sort of pre-programmed design.

strudel — a product prepared by rolling paper-thin flour dough into cylinders containing fruit or other fillings, then baking.

strudel leaves — squares of unbaked very thin flour and water dough; some versions include a small proportion of whole eggs.

stuffed straight — a flour consisting of a straight flour to which has been added clears from another milling operation.

sublimation — emitting of vapor from a frozen (i.e., solid, not liquid) material; the way water is lost from ice in frozen products.

succinylated monoglyceride — a surfactant sometimes used in doughs to improve handling properties, enhance crumb texture, and lengthen shelf life.

sucre en poudre — *Fr* powdered sugar.

sucre glace — *Fr* confectioners', powdered, or 10X sugar.

sucre semoule — *Fr* sucrose in large particles, often used in making fruit preserves.

sucrose — the chemical substance of which common table sugar, beet or cane sugar, is about a 99% pure example. When hydrolyzed by invertase or mild acid treatment, sucrose yields the hexoses, fructose and glucose.

sucrose esters — sucrose molecules combined with 1, 2, or 3 fatty acids. Has been recommended as a dough strengthener and crumb softener.

sugar — when not modified, means table sugar — cane or beet sucrose. In the most general sense, means any carbohydrate having a sweet taste.

sugar batter procedure — a method of mixing cake batters that requires creaming the shortening with sugar, then blending in the eggs, colors, and flavors, and finally mixing this with flour and leaveners.

sugar wafers — cookies and cookie bases made by baking an unleavened batter of very low viscosity between two heated plates bearing a design of some sort, then interleaving two or more of these bases with a "creme" consisting mainly of a fatty material and finely ground sugar. Both the separate baked product and the finished creme cookie have been called "sugar wafers."

suhkrut — *Estonia* sugar.
suji — *Ind* semolina.
suji beveca — *Ind* a sweet dessert made from semolina.
sultanas — a type of raisin, common term in the U.K., not in U.S.
sunflower oil — oil that has been pressed or solvent-extracted from seeds of the sunflower. Contains a high percentage of polyunsaturated fats (around 69%).
superheated steam — steam that does not contain as much water could be held in a similar volume at equal temperature and pressure.
surfactants — chemical substances capable of reducing the surface tension of water. Important for their ability to improve emulsification of water and oil. Useful in baked products as means of increasing the "wettability" of water.
sweating — the changes in wheat induced by moisture and heat, as in tempering or during storage. Principally due to biological changes, such as respiration, but also may include chemical and physical processes.
Swiss roll — *UK* a jelly-roll type of confection consisting of a layer of baked cake coated with jam, jelly, marshmallow, or other filling, then rolled up in spiral fashion. Usually enrobed with chocolate couverture.
symmetry — the extent to which a loaf is uniformly proportioned.
syneresis — the separation of gels into a liquid portion and a shrunken solid (plastic) portion. Occurs with virtually all food gels, given sufficient time.
synergist — a substance that, when used in combination with another material, results in a combination having a greater effect (of some sort) than would be predicted from the sum of the effects of the individual materials.
synthetic — a substance made artificially by combining (usually by chemical reactions) other materials; not appropriately applied to simple mixtures, however.

-T-

taco — *Mex* a tortilla wrapped around some sort of savory filling; dozens if not hundreds of major variations exist, such as taquitos (smaller diameter), flautas (still smaller), enchiladas (covered with sauce), etc.

taftoon — *Iran* round, sourdough flat bread with small holes on its surface.

tailings — material that is too coarse to pass through the screens in a bolter or sifter.

tail of the mill — the part of the mill where the final processing occurs.

tail of the mill particles — wheat particles of relatively small size that are removed near the end of the milling plant.

tail sheet — a coarser screen at the far end of a sifter, intended to scalp stock that is coarser than the material screened out by the rest of the sieves.

tainas — *Estonia* dough.

tallow — hard beef fat obtained by rendering fatty tissues removed during butchering; sometimes also applied to fat from sheep, etc. There are edible and inedible forms, the latter mostly used as a raw material for soap. Tallow contains a mixture of oils and hard fats of potential value as shortening ingredients, e.g., oleo oil and stearine.

talmouse — *Fr* cheesecake.

tamale — masa spread thinly on a corn husk, partially covered with comminuted or shredded meat and sauce, then rolled up to make a cylinder with meat in the middle. The husk holds the assembly together during further cooking and is removed before the tamale is eaten.

tamatar — *Ind* tomato.

tamees — *Saudi Arabia* thin flatbread made from a very lean yeast-leavened dough.

tandouri roti — *Pakistan* a flat bread, the preparation method and formula for which are similar to those used for chapati, but baked in a tandoori (tanduri), an oval in-ground oven having its walls plastered with clay.

tannouri — *Saudi Arabia* round flatbread made from a lean yeast-leavened dough that undergoes one fermentation and one proofing stage. Has a dark-spotted, golden brown crust, little or no crumb, and a docked upper surface.

tanok — *Iran* round flatbreads baked in the countryside from water, salt, flour, and starter. Vary in thickness from 3.5 cm to paper thin.

tapioca — a food material (largely starch) obtained from roots of cassava plants. At one time, it was widely used to prepare puddings in home kitchens. It can be further processed to yield an industrial starch with some unusual properties.

tap water — generally means potable water delivered by a municipal supplier through their piping system.

tartaric acid — an acid obtained as a by-product of wine preparation, and from other sources. Used as a mild acidifier in food products; was at one time a common acidulant in baking powders.

tarteletter — *Denmark* patty shells.

tarts — small pies, usually sized for individual servings.

taste — the sensations sweet, sour, salt, and bitter. Some authorities claim it should include other mouth and tongue sensing capabilities, e.g., chemical heat sensations (as from pepper). Not properly used to include aromatic properties, however.

TBHQ — an antioxidant effective in retarding development of oxidative rancidity.

tea rolls — small sweet buns.

tear strength — a test applied under standardized conditions to give a measure of the resistance of packaging film to tearing after a hole or slit has been made in the film.

Teflon — a white waxy material that has low friction and very low adhesive properties; used for preventing the sticking of doughs and the like to processer surfaces.

tejolote — *Mex* a pestle used with a molcajete to grind nixtamal for tortillas; usually wide at bottom and narrow at top, like a typical lab pestle.

tempering — similar in meaning to conditioning, but applied occasionally to materials other than grain. Also, can mean allowing an ingredient or product to come to the same temperature as the space in which it is stored.

tempura ko — *Jap* flour used to make the tempura batter for coating shrimp and raw vegetables that are to be deep-fried. Often, a low-gluten wheat flour.

tenderness — used by bakers to describe the fragility or physical strength of the crumb or crust of a baked product.

tensile strength — the greatest longitudinal stress a substance can bear without tearing apart, usually expressed with reference to a unit area of cross-section.

terabelesi — *Tunisia* round flatbread made from a flour, water, salt, and yeast dough that is fermented and proofed. Cuts forming a square are made across the top of the dough piece before it is baked.

terminal velocity — rate of aspiration or air flow that is just sufficient to lift a given type of particle from a mixture, usually reported in ft per min.

tészták — *Hungary* dumplings.

texture — the mouthfeel, consistency, or other effects a product has on the tactile senses, including those involved in chewing. When applied to bread, it usually means only the sensation obtained when the fingers of the observer are lightly passed across the crumb surface. Terms typically used are smooth, silky, harsh, etc.

thermoduric — describes microorganisms that can withstand high temperatures.

thermostat — the combination of sensing and activating mechanisms used to maintain the temperature of a gas or liquid within a narrow range by modulating a heating and/or cooling source.

thiamin — a B vitamin, one of those that must be added to enriched flour.

throughs — particles that pass through a bolter or sieve.

tight — a baker's term for a stiff, tough, or bucky dough.

tikerberikook — *Estonia* gooseberry cake.

til — *Ind* sesame seeds.

tin bread — *UK* pan bread.

tines — (1) The prongs of a fork or the like. (2) Pointed metal rods used in equipment for fork-splitting English muffins.

titanium dioxide — a non-certified color, insoluble in water, that is useful in whitening icings and the like.

tomate — *Sp* tomato.

top crust — the part of the crust that has not been in contact with the pan during baking.

top heat — the heat a product receives from the oven parts above the product.

topping flour — flour added at the dough stage, in the sponge-and-dough process for making bread.

tortelettes — *Fr* small round cakes.

torten — or tortes. Cakes of the continental type, especially large fancy cakes enriched with creams, marzipans, chocolate, fruits, and nuts; also used for pastries that resemble coffee cakes or pies.

tortilla — traditionally, a thin, disc-shaped, unleavened cake hand-formed from ground lime-treated partly gelatinized corn kernels and surface baked (as on a griddle). There are now versions made from wheat flour that often include a slight amount of baking powder. Tortillas are the basis for many foods in Mexican cuisine, e.g., tacos, burritos, and enchiladas.

tortilla chips — pieces cut or torn from tortillas and fried. Also, baked or fried extruded pieces of masa or corn meal resembling the traditional product in one or more of their characteristics.

tortillina — *Mex* a word sometimes used to indicate tortillas made from wheat flour.

tort makowy — *Poland* poppy seed torte.

tort piaskowy — *Poland* sand cake.

total dietary fiber — all of the various non-digestible vegetable components, as determined by various tests, not all of them accurate or reliable.

Total Fat — according to 1993 labeling regulations, the sum of fatty acids from mono-, di-, and tri-glycerides, free fatty acids, phospholipid fatty acids, and sterol fatty acids expressed as triglycerides.

total titratable acidity — the total acid produced by fermentation, as determined by direct titration with standardized alkali solution.

Totox Value — a determination of the degree of fat oxidation. The following formula is used: Totox = 2 (peroxide value) + anisidine value.

tough — a non-specific term used to describe the resistance of dough to stretching or tearing.

tragacanth — a natural gum soluble in hot water that has been used as a component of glazes, as a binder in confectionery, and as a stabilizer for ice cream.

tramp iron — bits of metal found in the grain as it comes to the mill.

translucent — not opaque but not entirely transparent; describes materials that transmit light with considerable internal dimming and distortion.

transparent — (1) Allowing the passage of radiation without substantially reducing its energy; in its widest sense applicable to both visible and invisible radiation. (2) The property of a material which allows observation of objects through intervening thicknesses of the material without substantial distortion or blurring, as in the case of sheets of clear glass.

transpositor — the mechanism in overhead proofers that inverts dough pieces about halfway through the proofing period.

traveling hearth ovens — ovens in which the hearth consists of steel plates linked together and mechanically drawn through the baking chamber. Straps of pans are placed on the hearth at one end of the chamber and the baked product removed at the other end.

traveling tray ovens — ovens in which trays are pulled along a horizontal track by chains; the trays are loaded with panned dough pieces at the front end, travel from front to rear of the chamber, are transferred to a lower track, and are then carried back to the starting point (but at a lower level) where the baked product is unloaded.

trays — plastic containers into which bakery products are placed for transfer from the production point to sales outlets.

trifle — a multiple-serving dessert traditionally made from a combination of cake, wine, macaroons, and fruit preserves. A typical modern example might consist of liqueur-soaked ladyfingers positioned around the inside of a glass bowl which is then filled with Bavarian cream; this assembly is sometimes topped with whipped cream and/or it may contain jelly or fruit preserves in a recognizable layer. Some trifle recipes are very elaborate.

triglycerides — chemical esters formed by combining one unit of glycerol with three units of fatty acids.

trigo — *Sp* wheat.

trigo sarraceno — *Sp* buckwheat.

tristimulus — a method of specifying color that relies on the quantification of three factors: e.g., hue, brilliance, and saturation. Other systems have also been called "tristimulus."

triticale — a plant that combines the genetic material of wheat and rye; the grain can be milled into a flour that makes fairly good bread.

trolley cookies — fairly soft base cakes that are dipped in successive coats of, e.g., marshmallow, icing, and chocolate couverture, by a device consisting of pins (to hold the cookies) mounted on an elaborate conveying device (the trolley).

trough — a large container of U-shaped cross section, usually on casters, for holding batches of dough during bulk fermentation and for transporting the dough between processing areas.

trough hoist — a device for raising a dough trough above the mixer so the sponge can be easily transferred to the bowl.

tsoureki — Greek Easter bread, rich yeast-leavened dough formed into three cylinders then plaited; usually doesn't include fruits, but may include one hard-boiled egg in its (colored) shell in each large pastry.

tunnel ovens — ovens with baking hearths made of steel segments that move through the chamber; they are loaded at one end and discharge baked goods at the other end. Customarily used as one-product ovens in large wholesale bakeries.

turbinado sugar — raw sugar that has been partially refined by washing in a centrifuge to remove molasses from the surface of the granules. Darker and less purified than regular cane or beet sugar.

turn — in the production of puff pastry dough, the operation of folding the dough in layers after each sheeting operation, usually after turning the assembly 90 degrees.

turnover — (1) The rate at which fat is used up in a frying operation. (2) A single-serving dessert product made of a disc or square of puff pastry dough, one-half of which is covered with a fruit (or other) filling, the uncovered half being folded over to conceal the filling, and the ensemble then baked.

turntable — a rotatable cake stand used by decorators to simplify application of icing and the like to a cake.

túrós töltelék — *Hungary* cottage cheese filling, as for pancakes.

tutti frutti — *It* mixed glazed fruits.

twist — designates a type of weave used in fabrics for sieves. In half-twist cloth, every alternate warp consists of two half-size threads, one passing over the woof and the other under it. In full-twist, every warp consists of two half-size threads.

twist bread — a method of molding bread that consists of forming two cylinders, each half the desired weight of the panned loaf, then twisting them around each other before they are placed in the pan. Believed to improve the appearance and texture of the finished loaf.

twister — the production machine that joins the two or more cylinders of twist-bread dough to form a finished raw loaf.

-U-

udon — *Jap* noodles made of wheat flour.

ultraviolet light — the short wave lengths of electromagnetic radiation occuring just beyond the blue or violet end of the visible spectrum. UV radiation has some sterilizing capabilities.

uncertified color additive — color additive approved for food use but not listed as a certified color. Examples are annatto extract, beet powder, and titanium dioxide.

unleavened — a dough or batter that does not contain an ingredient added for the purpose of generating gas within the mixture; it is the usual practice to describe doughs or batters that do not contain yeast, baking powder, ammonium bicarbonate, etc., as "unleavened" even though they may expand considerably in the oven as the result of the increased internal pressure of water vapor and air.

unsaponifiable matter — substances found in natural fats and oils that cannot be saponified, i.e., do not split when reacted with aqueous alkalies. Includes some of the more complex organic chemical compounds that are commonly soluble in ordinary fat solvents.

unsaturated — (1) A liquid not containing the maximum amount of a substance that would be soluble in it under the conditions of observation. (2) A term descriptive of the distribution of hydrogen atoms on the carbon "backbone" of a fatty acid (chemically, the term has a much broader meaning). In a fully saturated molecule, each of the carbon atoms (except the end atoms) have two hydrogen atoms attached. If the molecule is mono-unsaturated, two adjacent carbon atoms in the molecule will each have only one hydrogen atom attached. In polyunsaturated fats, more than one pair of adjacent carbon atoms will lack the full complement of hydrogen atoms.

unsulfured molasses — molasses that has not been treated with sulfur dioxide during its processing; usually applied to the so-called whole cane juice molasses, a premium grade.

-V-

vacuum cooler — a chamber in which hot bread loaves are placed, there being subjected to a partial vacuum so that the loaves cool by the evaporation of some of the water they contain.

vainilla — *Mex* vanilla.

vanilla — the flavor prepared from cured vanilla beans, usually by extracting them with ethanol solutions. Imitation vanilla flavor consists of vanillin with other synthetic and natural materials such as ethyl vanillin.

vanilla beans — the fruit of a tropical orchid. The long pods are put through a "curing" process, during which enzymic and other changes lead to development of the typical vanilla flavor.

vanillin — a chemical substance, 4-hydroxy-3-methoxy-benzaldehyde, responsible for the top note of both natural and artificial vanilla aroma.

vegetable colors — materials obtained from plants, used for coloring foods; beet powder is a typical example. Commercial vegetable colors have all been purified and concentrated to a greater or lesser degree.

velocity — speed at which a solid, liquid, or gas is moving; rate of motion, distance moved in unit time.

ventilate — to increase the movement of ambient air through a space, e.g., by making openings in an enclosure to facilitate natural air currents or by using fans.

vertical mixer — any mixer having an agitator suspended in a vertical position; planetary mixers are all vertical mixers, but not all vertical mixers are planetary mixers.

vett — *Estonia* water.

Vienna bread — a hearth-type yeast-leavened bread in loaf form with heavy crisp crust, made from a lean dough. The loaf is usually elongated and somewhat pointed at the ends, and the top may be slashed (usually lengthwise.)

Viennoisierie — *Fr* Vienna-type bakery products.

vinegar — a dilute (usually 5%) solution of acetic acid produced either by microbiolgical action on fermentation alcohol (cider, malt, or wine vinegar) or by chemically oxidizing ethanol from any source (white or distilled vinegar).

vinagre — *Sp* vinegar.

violets — candied violet flowers were formerly used as decorations for fancy baked products, but have been replaced by artificial blossoms made from colored sugar icing.

viscometer — any of a number of instruments designed to measure the viscosity or apparent viscosity of fluids.

viscosity — a liquid's resistance to flow.

viscous — having a relatively high resistance to flow.

vital wheat gluten — purified gluten prepared from wheat in such a manner that many of the native properties (including extensibility) are retained; contrasted to denatured wheat gluten, which will not form a cohesive extensible mass when rehydrated.

vitamins — organic substances required in trace amounts for catalyzing or otherwise supporting normal metabolic processes in living organisms and, usually, not synthesized by the organisms in amounts sufficient to promote optimum health. Different species may require different vitamins in their diet.

vitreous — glassy. As one example, used to describe the cut surface of a kernel of hard wheat, when it refers to a somewhat shiny, slick, translucent appearance.

vol-au-vent — a circular puff pastry shell with its center open at the top, like a cup. Either used for entrees like creamed chicken or filled with fruit and cream.

votator — a device for plasticizing fats, basically a scraped surface heat exchanger through which the melted fats are pumped. Heat is removed from the fat until crystallization or solidification of some of the fractions occurs. The solidified or plastic material exiting the votator is of homogeneous composition and of a consistency permitting it to be molded or packaged into cartons as though it were a solid product. The material also gives different results (usually) than the liquid oil constituents when it is used as an ingredient in cakes and cookies.

-W-

wafer — (1) A thin cookie. (2) A thin tablet or pellet containing a precise amount of nutrient additives or dough modifiers, often used for supplementing bread doughs.

waffeln — *Ger* sugar wafers.

waffle — a firm, fairly thick cake with deep, regular surface indentations, baked in a special mold. Somewhat similar to a pancake in composition and usage although it is considerably lower in moisture content than a regular pancake and it frequently contains some eggs and milk.

walnut — (1) The English walnut, *Juglans regia*. (2) The black walnut, *Juglans nigra*.

wash — a liquid brushed on the surface of a baked or unbaked product to alter its crust characteristics (appearance, texture, flavor, etc.).

water absorption — the percentage of water, relative to flour as 100%, required to give a dough of the desired consistency.

water activity — ratio of the relative humidity of the atmosphere in equilibrium with a water-binding substance as compared to the relative humidity of the atmosphere above pure water held at the same temperature and pressure.

water splitter — a device that uses a high velocity jet of water to cut a slit in the top of bread dough.

wax — a relatively tough, meltable, solid material insoluble in water; a typical example is beeswax. Synthetic and natural kinds are available. Waxes are generally not digestible.

waxed paper — thin paper coated on one or both sides with petroleum wax, sometimes supplemented with plastic resins. Once widely used for packaging bread loaves, but now rarely so used.

waxy maize — a variety of corn having nearly all of its starch present in the form of branched molecules (amylopectin).

weizenbrötchen — *Ger* bread rolls of various types, yeast-leavened.

weizenmischbrot — *Ger* the German standard bread consisting of a mixture of more than 50% wheat flour with a lesser amount of rye flour. Given a relatively short fermentation.

Westphalian pumpernickel — the darkest, densest, dankest kind of pumpernickel; seldom made in the U.S.A.

wet milling — a process for separating corn kernels into their component parts using a water-sulfur dioxide system in combination with milling and separation equipment.

wet peak — a stage in the egg whipping process; when the mixing utensil is withdrawn from the meringue (for example) the peak of the mixture will glisten with free liquid and will, usually, soon fall over.

GLOSSARY OF MILLING AND BAKING TERMS

wheat — the plant of the genus *Triticum*, and its seed.

wheat bread — bread made from a dough containing both whole wheat flour and white flour.

wheat germ — the germ or embryo of the wheat seed; contains a fairly large amount of oil and oil-soluble vitamins. Also contains substances that weaken gluten.

whey — the residue of milk after the butterfat and curds have been removed in the preparation of cheese; consists principally of water, albumin proteins, inorganic salts, and lactose.

whipped cream — cream of high butterfat content that has been whipped or beaten to make it incorporate large amounts of air, or that is dispensed from a pressurized container. Usually contains sugar and vanilla flavor.

whipping — beating a fluid or semi-fluid material such as egg whites or cream to incorporate air and then to subdivide the bubbles into a fine and uniform foam.

white wheat — a soft wheat with a light-colored hull and a relatively friable endosperm; suitable for pastry flours. Grown in relatively small quantities in the U.S.A.

whizzing — whirling grain in a special centrifuge to rid it of surplus water that has been added during a washing process.

whole wheat bread — bread made from dough that has its entire wheat content as whole grain meal or flour.

wicket — a simple metal attachment for holding the stack of plastic bags used in automatic bread bagging machines.

wicket hole — the small holes punched in a plastic bread bag to permit it to be threaded on to the wicket.

wicket pack — the wicket with its complement of 100 or 200 plastic bread bags, as delivered by the supplier.

wiener — (1) A baked product allegedly of Viennese or Austrian type or origin. (2) A sausage of the hot dog shape.

wienerbrodsdejg — *Denmark* dough for Danish pastry (or, as they call it, Vienna bread).

wild break — unusually large break and shred at the side of the baked loaf, often irregular. This fault is not only unsightly, but it suggests the likelihood that other defects will be found in the bread's texture and uniformity.

wild yeast — basically, any yeast other than *Saccharomyces cerevisiae* that unexpectedly contributes to leavening and flavor development in a dough. The changes are nearly always for the worse.

Wiley melting point — a specialized test for fats that measures the liquefaction of a disc of fat heated under certain conditions.

wine — fermented fruit juice; commercial versions often contain additives that are not declared on the container. If made from any fruit except grapes, a modifier must be added to the name, e.g., peach wine.

winterization — a process applied to food oils to remove naturally occurring high-melting triglycerides so that the oil will not become cloudy during cold storage. The oil is held at a reduced temperature until the crystallization of high melting point fats is complete, and then the solidified fats are filtered out.

winter wheat — wheat that is sown in the late fall or early winter and harvested in the late spring or early summer; most U.S.A. wheat is of this type.

wire-cut cookies — the dough for these cookies is extruded as a continuous cylinder through an orifice; the dough strand is cut cross-wise by a wire (or band) drawn across the orifice at regular intervals. The slices of dough so formed fall directly on the oven band. Various cross-sectional shapes are possible.

wort — the liquid mixture prepared for fermentation in the brewing process or as an intermediate product in the manufacture of malt syrup.

-X-

xanthan gum — a gum produced by the action of the microorganism *Xanthomonas campestris* on glucose. When combined with locust bean gum, a synergistic reaction occurs that leads to gel formation.

x-ray crystallography — a method for determining the physical arrangement of atoms or molecules in crystalline materials. It can be used for studying the polymorphic form of a solidified fat, for example.

-Y-

yeast — the unicellular microorganism that is responsible for leavening bread, etc. Also, commercial preparations consisting mostly of these cells.

yeast food — a mixture designed to be added to a dough to increase yeast activity and, particularly, to improve cell multiplication. Many of these proprietary products consist mainly of mineral salts.

yeast substrate — a substance used by yeast for its metabolic processes, including those, like sugar, transformed by fermentation into alcohol and carbon dioxide.

yield — in general, the amount of any product or intermediate resulting from the processing of a given amount of ingredients or intermediate products, expressed either as a percentage of the beginning materials or as final weight. Specific examples are: the amount of finished mill products expressed as a percentage of the amount of grain required to make them, and the weight of baked product resulting from 100 pounds of ingredients.

yit bien — *Chi* moon cake, a baked bun with filling; chemically leavened.

yogurt — milk thickened by acids that have been developed in the fluid through bacterial action. Several other spellings will be found, e.g., yoghurt. Modern consumer types are nearly always sweetened and flavored with additives. Has been used as an ingredient in doughs, batters, and fillings.

young — describes a yeast dough that is underfermented. These doughs produce baked products that are light in color, tight in grain, and low in specific volume.

yufka — *Turkey* this name has evidently been applied to more than one kind of product, but a common form is a thin, round flatbread made from an unleavened dough.

-Z-

zest — finely grated peels of citrus fruits.

zweiback — see "zwieback."

zwieback — this word is so often misspelled "zweiback" that the latter version should perhaps be regarded as having developed into an acceptable alternative spelling (at least in the U.S.A.). Zwieback is a kind of rich bread or coffee cake that is first baked in loaf form (often with a circular cross-section) and then sliced and toasted to a low moisture content in a manner that allows only slight to moderate browning. Typically, the texture is crisp, the color light brown, and the flavor fairly bland.

zymase — a combination of enzymes (as from yeast) that, acting in concert, transform certain sugars into carbon dioxide and ethanol.